Conversations with Stalin

By Milovan Djilas

THE NEW CLASS
LAND WITHOUT JUSTICE
ANATOMY OF A MORAL
CONVERSATIONS WITH STALIN

Conversations
with
Stalin

Milovan Djilas

Translated from the Serbo-Croat by
Michael B. Petrovich

Harcourt, Brace & World, Inc., New York

To the memory of
ANEURIN BEVAN

Contents

Note on the Spelling and Pronunciation of
Serbo-Croat Words and Names

s = s as in sink
š = sh as in shift
c = ts as in mats
č = ch as in charge
ć = similar to, but lighter than, č—as in arch
ž = j as in French *jour*
z = z as in zodiac
j = y as in yell
nj = n as in neutral
g = g as in go
dj = g as in George
lj = li as in million

Conversations with Stalin

Foreword

It is in the nature of the human memory to rid itself of the superfluous, to retain only what has proved to be most important in the light of later events. Yet that is also its weak side. Being biased it cannot help adjusting past reality to fit present needs and future hopes.

Aware of this, I have endeavored to present the facts as exactly as possible. If this book is still not exempt from my views of today, this should be attributed neither to ill will nor to the partisanship of a protagonist, but rather to the nature of memory itself and to my effort to elucidate past encounters and events on the basis of my present insights.

There is not much in this book that the well-versed reader will not already know from published memoirs and other literature. However, since an event becomes more comprehensible and tangible if explained in greater detail and from several vantage points, I have considered it not unuseful if I, too, had my say. I hold that humans and human relationships are more important than dry facts, and so I have paid greater attention to the former. And if the book contains anything that might be called literary, this too should be ascribed less to my style of expression than to my desire to make the subject all the more engaging, clear, and true.

While working on my autobiography, the idea occurred to me, in 1955 or 1956, to set apart my meetings with Stalin in a separate book which could be published sooner and separately. However, I landed in jail, and it was not convenient for me, while imprisoned, to engage in that kind of literary activity since, even though my book dealt with the past, it could not but impinge on current political relations.

Only upon my release from prison, in January of 1961, did I return to my old idea. To be sure, this time, in view of changed conditions and the evolution of my own views, I had to approach this subject rather differently. For one thing, I now devoted greater attention to the psychological, the human aspects of these historical events. Moreover, accounts of Stalin are still so contradictory, and his image is still so vivid, that I have also felt it necessary to present at the end, on the basis of personal insights and experiences, my own conclusions about this truly enigmatic personality.

Above all else, I am driven by an inner necessity to leave nothing unsaid that might be of significance to those who write history, and especially to those who strive for a freer human existence. In any case, both the reader and I should be satisfied if the truth is left unscathed, even if it is enveloped in my own passions and judgments. For we must realize that, however complete, the truth about humans and human relations can never be anything but the truth about particular persons, persons in a given time.

Belgrade
November 1961

I
Raptures

T HE first foreign military mission to come to the Supreme Command of the Army of People's Liberation and Partisan Units of Yugoslavia was the British. It parachuted in during May 1943. The Soviet Mission arrived nine months later—in February 1944.

Soon following the arrival of the Soviet Mission the question arose of sending a Yugoslav military mission to Moscow, especially since a mission of this kind had already been assigned to the corresponding British Command. In the Supreme Command, that is, among the members of the Central Committee of the Communist Party of Yugoslavia who were working at headquarters at the time, there developed a fervent desire to send a mission to Moscow. I believe that Tito brought it orally to the attention of the Chief of the Soviet Mission, General Korneev; however, it is quite certain that the matter was settled by a telegram from the Soviet Government. The sending of a mission to Moscow was of manifold significance to the Yugoslavs, and the mission itself was of a different character and quite different purpose to the one assigned to the British Command.

As is known, it was the Communist Party of Yugoslavia that organized the Partisan and insurgent movement against the German and Italian forces of occupation in Yugoslavia and their domestic collaborators. While solving its national problems through the most ruthless kind of warfare, it continued to regard itself as a member of the world Communist movement, as something inseparable from the Soviet Union—"the homeland of socialism." Throughout the entire war the innermost agency of the Party, the Political Bureau, more popularly known by the abbreviated name Politburo, managed to keep a connection with Moscow by radio. Formally this connection was with the Communist International—the Comintern—but at the same time it meant a connection with the Soviet Government as well.

The special conditions brought on by war and the survival of the revolutionary movement had already, on several occasions, led to misunderstandings with Moscow. Among the most significant I would mention the following.

Moscow could never quite understand the realities of the revolution in Yugoslavia, that is, the fact that in Yugoslavia simultaneously with the resistance to the forces of occupation a domestic revolution was also going on. The basis for this misconception was the Soviet Government's fear that the Western Allies, primarily Great Britain, might resent its taking advantage of the misfortunes of war in the occupied countries to spread revolution and its Communist influence. As is often the case with new phenomena, the struggle of the Yugoslav Communists was not in accord with the settled views and indisputable interests of the Soviet Government and state.

Nor did Moscow comprehend the peculiarities of warfare in Yugoslavia. No matter how much the struggle of the Yugoslavs enheartened not only the military—who were fighting to preserve the Russian national organism from the Nazi German invasion—but official Soviet circles as well, the latter nevertheless underrated it, if only by comparing it with their own Partisans and their own methods of warfare. The Partisans in the Soviet Union were an auxiliary, quite incidental force of the Red Army, and they never grew into a regular army. On the basis of their own experience, the Soviet leaders could not comprehend that the Yugoslav Partisans were capable of turning into an army and a government, and that in time they would develop an identity and interests which differed from the Soviet—in short, their own pattern of life.

In this connection one incident stands out as extremely significant to me, perhaps even decisive: In the course of the so-called Fourth Offensive, in March of 1943, a parley between the Supreme Command and the German commands took place. The occasion for the parley was an exchange of prisoners, but its essence lay in getting the Germans to recognize the rights of the Partisans as combatants so that the killing of each other's wounded and prisoners might be halted. This came at a time when the Supreme Command, the bulk of the revolutionary army, and thousands of our wounded found themselves in mortal danger, and we needed every break we could get. Moscow had to be informed about all this, but we knew full well—Tito because he knew Moscow, and Ranković more by instinct—that it was better not to tell Moscow everything. Moscow was simply informed that we were negotiating with the Germans for the exchange of the wounded. However, in Moscow they did not even try to project themselves into our situation, but doubted us—despite the rivers of blood we had already shed—and replied very sharply. I remember—it was in a mill by the Rama River on the eve of our breakthrough across the Neretva, February 1943—how Tito reacted to all this: "Our first duty is to look after our own army and our own people."

This was the first time that anyone on the Central Committee openly formulated our disparateness to Moscow. It was also the first time that my own consciousness was struck, independently of Tito's words but not unrelatedly, that this disparateness was essential if we wanted to survive in this life-and-death struggle between opposing worlds.

Still another example occurred on November 29, 1943, in Jajce, at the Second Session of the Antifascist Council, where resolutions were passed that in fact amounted to the legalization of a new social and political order in Yugoslavia. At the same time there was formed a National Committee to act as the provisional government of Yugoslavia. During the preparation for these resolutions in meetings of the Central Committee of the Communist Party, the stand was taken that Moscow should not be informed until after it was all over. We knew from previous experience with Moscow and from its line of propaganda that it would not be capable of understanding. And indeed, Moscow's reactions to these resolutions were negative to such a degree that some parts were not even broadcast by the radio station Free Yugoslavia, which was located in the Soviet Union to serve the needs of the resistance movement in Yugoslavia. Thus the Soviet Government failed to understand the most important act of the Yugoslav revolution—the one that transformed this revolution into a new order and brought it onto the international scene. Only when it became obvious that the West had reacted to the resolutions at Jajce with understanding did Moscow alter its stand to conform with the realities.

Yet the Yugoslav Communists, despite all their bitterness over experiences whose significance they could comprehend only after the break with Moscow in 1948, and despite their differing ways of life, considered themselves to be ideologically bound to Moscow and regarded themselves as Moscow's most consistent followers. Though vital revolutionary and other realities were separating the Yugoslav Communists ever more thoroughly and irreconcilably from Moscow, they regarded these very realities, especially their own successes in the revolution, as proofs of their ties with Moscow and with the ideological programs that it prescribed. For the Yugoslavs, Moscow was not only a political and spiritual center but the realization of an abstract ideal—the "classless society"—something that not only made their sacrifice and suffering easy and sweet, but that justified their very existence in their own eyes.

The Yugoslav Communist Party was not only as ideologically unified as the Soviet, but faithfulness to Soviet leadership was one of the essential elements of its development and its activity. Stalin was not only the undisputed leader of genius, he was the incarnation of the very idea and dream of the new society. That idolatry of Stalin's personality, as well as of more or less everything in the Soviet Union, acquired irrational forms and proportions. Every action of the Soviet Government—for example, the attack on Finland—and every negative feature in the Soviet Union—for example, the trials and the purges—were defended and justified. What appears even stranger, Communists succeeded in convincing themselves of the propriety and suitability of such actions, and in banishing from their minds unpleasant facts.

Among us Communists there were men with a developed aesthetic sense and a considerable acquaintance with literature and philosophy, and yet we waxed enthusiastic not only over Stalin's views but also over the "perfection" of their formulation. I myself referred many times in discussions to the crystal clarity of his style, the penetration of his logic, and the harmony of his commentaries, as though they were expressions of the most exalted wisdom. But it would not have been difficult for me, even then, to detect in any other author of the same qualities that his style was colorless, meager, and an unblended jumble of vulgar journalism and the Bible. Sometimes the idolatry acquired ridiculous proportions: we seriously believed that the war would end in 1942, because Stalin said so, and when this failed to happen, the prophecy was forgotten—and the prophet lost none of his superhuman power. In actual fact, what happened to the Yugoslav Communists is what has happened to all throughout the long history of man who have ever subordinated their individual fate and the fate of mankind exclusively to one idea: unconsciously they described the Soviet Union and Stalin in terms required by their own struggle and its justification.

The Yugoslav Military Mission went to Moscow, accordingly, with idealized images of the Soviet Government and the Soviet Union on the one hand and with their own practical needs on the other. Superficially it resembled the mission that had been sent to the British, but in composition and conception it in fact marked an informal nexus with a political leadership of identical views and aims. More simply: the Mission had to have both a military and a Party character.

2

Thus it was no accident that, in company with General Velimir Terzić, Tito assigned me to the Mission in my role as a high Party functionary. (I had by then been a member of the innermost Party leadership for several years.) The other members of the Mission were similarly selected as Party or military functionaries, and among them was one financial expert. The Mission also included the atomic physicist Pavle Savić, with the aim of having him pursue his scientific work in Moscow. We also had with us Antun Augustinčić, a sculptor, who was given a respite from the rigors of the war so that he might pursue his art. All of us, to be sure, were in uniform. I had the rank of general. I believe that my selection was based in part on the fact that I knew Russian well—I had learned it in prison during the years before the war—and in part because I had never been to the Soviet Union before and thus was not burdened with any factional or deviationist past. Neither had the other members of the Mission ever been to the Soviet Union, but none of them had a good command of Russian.

It was the beginning of March 1944.

Several days were spent in assembling the members of the Mission and their gear. Our uniforms were old and motley, and since cloth was lacking, new ones had to be made from the uniforms of captured Italian officers. We also had to have passports in order to cross British and American territories, and so they were hastily printed. These were the first passports of the new Yugoslav state and bore Tito's personal signature.

The proposal arose almost spontaneously that gifts be sent to Stalin. But what kind and from where? The Supreme Command was located at the time in Drvar, in Bosnia. The immediate surroundings consisted almost entirely of gutted villages and pillaged, desolated little towns. Nevertheless a solution was found: to take Stalin one of the rifles manufactured in the Partisan factory in Užice in 1941. It was quite a job to find one. Then gifts began to come in from the villages—pouches, towels, peasant clothing and footwear. We selected the best among these—some sandals of untanned leather and other things that were just as poor and primitive. Precisely because they were of this character, we concluded that we ought to take them as tokens of popular good will.

The Mission had as an objective to arrange for Soviet help to the People's Liberation Army of Yugoslavia. At the same time Tito charged us with gaining, either through the Soviet Government or other channels, UNRRA aid for the liberated areas of Yugoslavia. We were to ask the Soviet Government for a loan of two hundred thousand dollars to cover the expenses of our missions in the West. Tito emphasized that we declare that we would repay the sum as well as the aid in arms and medicine when the country was liberated. The Mission had to take with it the archives of the Supreme Command and of the Central Committee of the Communist Party.

Most important of all, it had to sound out the Soviet Government on the possibility of their recognizing the National Committee as the provisional legal government and of having the Soviets influence the Western Allies in this direction. The Mission was to maintain communications with the Supreme Command through the Soviet Mission, and it could also make use of the old channel of the Comintern.

Besides these tasks of the Mission, Tito charged me at our leave-taking to find out from Dimitrov, or from Stalin if I could get to him, whether there was any dissatisfaction with the work of our Party. This command of Tito's was purely formal—to call attention to our disciplined relations with Moscow—for he was utterly convinced that the Communist Party of Yugoslavia had brilliantly passed the test, and uniquely so. There was also some discussion about the Yugoslav Party émigrés (Communists who had gone to Russia before the war). Tito's attitude was that we were not to become involved in mutual recriminations with these émigrés, especially if they had anything to do with Soviet agencies and officials. At the same time Tito emphasized that I ought to beware of secretaries, for there were all kinds, which I understood to mean that we were not only to guard an already traditional Party morality, but that we were to avoid anything that might endanger the reputation and distinction of the Yugoslav Party and of Yugoslav Communists.

My entire being quivered from the joyous anticipation of an imminent encounter with the Soviet Union, the land that was the first in history—I believed, with a belief more adamant than stone—to give meaning to the dream of visionaries, the resolve of warriors, and the suffering of martyrs, for I too had languished and suffered torture in prisons, I too had hated, I too had shed human blood, not sparing even that of my own brothers.

But there was also sorrow—at leaving my comrades in the midst of the battle and my country in a death struggle, one vast battlefield and smoldering ruin.

My parting with the Soviet Mission was more cordial than my encounters with it usually were. I embraced my comrades, who were as moved as I was, and set out for the improvised airfield near Bosanski Petrovac. We spent the whole day there inspecting the airfield and conversing with its staff, which already had the air and habits of a regular and established service, and with the peasants, who had already grown accustomed to the new regime and to the inevitability of its victory.

Recently British planes had been landing here regularly at night, but not in great numbers—at most, two or three in the course of a single night. They transported the wounded and occasional travelers and brought supplies, most frequently medicines. One plane had even brought a jeep not long before—a gift from the British Command to Tito. It was at this same airfield, a month earlier, at high noon, that the Soviet Military Mission had landed in a plane on skis. In view of the terrain and other circumstances, this was a real feat. It was also quite an unusual parade, in view of the rather sizable escort of British fighter planes.

I regarded the descent and subsequent take-off of my plane too as quite a feat: the plane had to fly low over jagged rocks in order to come in for a landing on the narrow and uneven ice and, then, take off again.

How sorrowful and sunken in darkness was my land! The mountains were pale with snow and gashed with black crevices, while the valleys were devoured by the gloom, not a glimmer of light to the very sea and across. Below there was war, more terrible than any before, and on a soil that was used to the tread and breath of war and rebellion. A people was at grips with the invader, while brothers slaughtered one another in even more bitter warfare. When would the lamps light up the villages and towns of my land again? Would it find joy and tranquillity after all this hatred and death?

Our first stop was Bari, in Italy, where there was a sizable base of Yugoslav Partisans—hospitals and warehouses, food and equipment. From there we flew toward Tunis. We had to travel circuitously because of the German bases on Crete and in Greece. We stopped in Malta on the way, as guests of the British Commander, and arrived in Tobruk for the night just in time to see the whole sky licked by a murky fire which rose from the ruddy rocky desert below.

The next day we arrived in Cairo. The British lodged us discreetly in a hotel and placed a car at our disposal.

The merchants and the help took us for Russians because of the five-pointed stars on our caps, but it was pleasant to learn, as soon as we fleetingly mentioned that we were Yugoslav or spoke Tito's name, that they knew of our struggle. In one shop we were also greeted in our tongue with profanity, which the salesgirl had innocently learned from émigré officers. A group of these same officers, carried away by the longing to fight and homesickness for their suffering land, declared themselves for Tito.

Upon learning that the chief of UNRRA, Lehman, was in Cairo, I requested the Soviet Minister to take me to him that I might present him with our requests. The American received me without delay, but coldly, declaring that our requests would be taken into consideration at the following meeting of UNRRA and that UNRRA dealt only with legal governments as a matter of principle.

My primitive and catechismal conception of Western capitalism as the irreconcilable enemy of all progress and of the small and oppressed found justification even in my first encounter with its representatives: I noted that Mr. Lehman received us lying down, for he had his leg in a cast and was obviously troubled by this and the heat, which I mistook for annoyance at our visit, while his Russian interpreter—a hairy giant of a man with crude features—was for me the very image of a badman from a cowboy movie. Yet I had no reason to be dissatisfied with this visit to the obliging Lehman; our request was submitted and we were promised that it would be considered.

We took advantage of our three-day sojourn in Cairo to see the historic sights, and because the first chief of the British Mission in Yugoslavia, Major Deakin, was staying in Cairo, we were also his guests at an intimate dinner. From Cairo we went to the British base at Habbaniya, near Baghdad. The British Command refused to drive us to Baghdad on the grounds that it was not quite safe, which we took for concealment of a colonial terrorism we thought to be no less drastic than the German occupation of our country. Instead of this, the British invited us to a sports event put on by their soldiers. We went, and had seats next to the Commander. We looked funny even to ourselves, let alone to the polite and easygoing English, trussed up as we were in belts and buttoned up to the Adam's apple.

We were accompanied by a major, a merry and good-hearted old fellow who kept apologizing for his poor knowledge of Russian—he had learned it at the time the British intervened at Archangel during the Russian Revolution. He was enthusiastic about the Russians (their delegations too had stopped at Habbaniya), not about their social system but about their simplicity and, above all, their ability to down huge glasses of vodka or whisky at one gulp "for Stalin, for Churchill!"

The Major spoke calmly, but not without pride, of battles with natives incited by German agents, and indeed, the hangars were riddled with bullets. In our doctrinaire way we could not understand how it was possible, much less rational, to sacrifice oneself "for imperialism" —for so we regarded the West's struggle—but to ourselves we marveled at the heroism and boldness of the British, who had ventured forth and triumphed in distant and torrid Asian deserts, so few in numbers and without hope of assistance. Though I was not capable at the time of deriving broad conclusions from this, it certainly contributed to my later realization that there did not exist a single ideal only, but that there were on our globe countless co-ordinate human systems.

We were suspicious of the British and held ourselves aloof from them. Our fears were made especially great because of our primitive notions about their espionage—the Intelligence Service. Our attitudes were a mixture of doctrinaire clichés, the influence of sensational literature, and the malaise of greenhorns in the great wide world.

Certainly these fears would not have been as great had it not been for those sacks filled with the archives of the Supreme Command, for they contained also telegrams between ourselves and the Comintern. We found it suspicious too that everywhere the British military authorities had shown no more interest in these sacks than if they had contained shoes or cans. To be sure, I kept them at my side throughout the trip, and to avoid being alone at night, Marko slept with me. He was a prewar Communist from Montenegro, simple but all the more brave and loyal for that.

It happened in Habbaniya one night that someone silently opened the door of my room. I was aroused even though the door did not creak. I spied the form of a native in the light of the moon, and, getting enmeshed in the mosquito net, I let out a shout and grabbed the revolver under my pillow. Marko sprang up (he slept fully clothed), but the stranger vanished. Most probably the native had lost his way or intended to steal something. But his insignificant appearance was sufficient to make us see the long arm of the British espionage in this, and we increased our already taut vigilance. We were very glad when, the next day, the British placed at our disposal a plane for Teheran.

The Teheran through which we moved about, from the Soviet Command to the Soviet Embassy, was already a piece of the Soviet Union. Soviet officers met us with an easy cordiality in which traditional Russian hospitality was mixed in equal measure with the solidarity of fighters for the same ideal in two different parts of the world. In the Soviet Embassy we were shown the round table at which the Teheran Conference had been seated, and also the upstairs room in which Roosevelt had stayed. There was nobody there now and all was as he had left it.

Finally a Soviet plane took us to the Soviet Union—the realization of our dreams and our hopes. The deeper we penetrated into its gray-green expanse, the more I was gripped by a new, hitherto hardly suspected emotion. It was as though I was returning to a primeval homeland, unknown but mine.

I was always alien to any Panslavic feelings, nor did I look upon Moscow's Panslavic ideas at that time as anything but a maneuver for mobilizing conservative forces against the German invasion. But this emotion of mine was something quite different and deeper, going even beyond the limits of my adherence to Communism. I recalled dimly how for three centuries Yugoslav visionaries and fighters, statesmen and sovereigns—especially the unfortunate prince-bishops of suffering Montenegro

—made pilgrimages to Russia and there sought understanding and salvation. Was I not traveling their path? And was this not the homeland of our ancestors, whom some unknown avalanche had deposited in the windswept Balkans? Russia had never understood the South Slavs and their aspirations; I was convinced that this was because Russia had been tsarist and feudal. But far more final was my faith that, at last, all the social and other reasons for disagreements between Moscow and other peoples had been removed. At that time I looked upon this as the realization of universal brotherhood. But also as my personal bond with the being of the prehistoric Slavic community. Was not this the homeland not only of my forebears but also of warriors who were dying for the final brotherhood of man and the final domination by man over things?

I became embodied in the surge of the Volga and limitless gray steppes and found my primeval self, filled with hitherto unknown inner urges. It occurred to me to kiss the Russian soil, the Soviet soil which I was treading, and I would have done it had it not seemed religious, and, moreover, theatrical.

In Baku we were met by a commanding general, a taciturn giant of a man made coarse by garrison life, war, and the service—the incarnation of a great war and a great land opposing a ravaging invasion. In his rough cordiality he was nonplused by our almost shy restraint: "What kind of people are these? They don't drink, they don't eat! We Russians eat well, drink even better, and fight best of all!"

Moscow was gloomy and somber and surprisingly full

of low buildings. But what significance could this have beside the reception prepared for us? Honors according to rank and a friendliness which was purposely restrained because of the Communist character of our struggle. What could compare with the grandeur of the war that we believed would be mankind's final trial and that was our very life and our destiny? Was not all pale and meaningless beside the reality that was present precisely here, in the Soviet land, indeed, a land that was also ours and mankind's, brought forth from a nightmare into a tranquil and joyous actuality?

3

They billeted us in the Red Army Center, the TsDKA, a kind of hotel for Soviet officers. The food and all other features were very good. They gave us a car with a chauffeur, Panov, a man well along in years, simple, and somewhat bent, but of independent views. There was also a liaison officer, Captain Kozovsky, a young and very handsome lad who was proud of his Cossack origin, all the more so inasmuch as the Cossacks had "washed away" their counterrevolutionary past in the present war. Thanks to him we were always sure, at any time, of obtaining tickets for the theater, the cinema, or anything else.

But we were not able to make any serious contact with the leading Soviet personages, though I requested to be received by V. M. Molotov, then Commissar for Foreign Affairs, and, if possible, by J. V. Stalin, the Prime Minister and Commander in Chief of the armed forces. All my circuitous attempts to present our requests and needs were in vain.

In all this no help was to be had from the Yugoslav Embassy, which was still royalist, though Ambassador Simić and his small staff had declared themselves for Marshal Tito. Formally respected, they were in fact more insignificant and accordingly more powerless than we.

Nor could we accomplish anything through the Yugoslav Party émigrés. They were few in number—decimated by purges. The most distinguished personality among them was Veljko Vlahović. We were the same age, both revolutionaries out of the revolutionary student movement of Belgrade University against the dictatorship of King Alexander. He was a veteran of the Spanish Civil War, while I was coming from an even more terrible war. He was a man of great personal integrity, highly educated and wise, though excessively disciplined and not independent in his views. He managed the radio station Free Yugoslavia, and his co-operation was valuable, but his connections did not go beyond Georgi Dimitrov, who, since the Comintern had been dissolved, shared with D. Z. Manuilsky the direction of the section of the Soviet Central Committee for foreign Communist parties. We were well fed and graciously received, but as far as the problems we had to present and to solve were concerned, we could make no headway whatsoever. To tell the truth, it must be stressed again that, except for this, we were received with extraordinary geniality and consideration. But it was not until a month following our arrival, when Stalin and Molotov received General Terzić and me and this was published in the press, that all the doors of the ponderous Soviet administration and of the rarefied heights of Soviet society were magically thrown open.

The Panslavic Committee, which had been created in the course of the war, was the first to arrange banquets and receptions for us. But one did not have to be a Communist to perceive not only the artificiality but also the hopelessness of this institution. Its activity was centered on public relations and propaganda, and even in this it was obviously limited. Besides, its aims were not very clear. The Committee was composed almost entirely of Communists from the Slavic countries—the émigrés in Moscow who were in fact alien to the idea of Panslavic reciprocity. All of them tacitly understood that it was a matter of resurrecting something long since outmoded, a transitional form meant to rally support around Communist Russia, or at least to paralyze anti-Soviet Panslavic currents.

The very leadership of the Committee was insignificant. Its President, General Gundorov, a man prematurely grown old in every respect and of limited views, was not a man one could talk to effectively even on the simplest questions of how Slavic solidarity could be achieved. The Committee's Secretary, Mochalov, was rather more authoritative simply by virtue of being closer to the Soviet security agencies—something that he concealed rather badly in his extravagant behavior. Both Gundorov and Mochalov were Red Army officers, but were among those who had proved to be unfit for the front. One could detect in them the suppressed dejection of men demoted to jobs that they did not consider their line. Only their secretary, Nazarova, a gap-toothed and excessively ingratiating woman, had anything resembling love for the suffering Slavs, though her activities too, as was later learned in Yugoslavia, were subordinated to Soviet security agencies.

In the Panslavic Committee headquarters one ate well, drank even more, and mostly just talked. Long and empty toasts were raised, not much different from one another, and certainly not as beautiful as those of tsarist times. I was truly struck by the absence of any freshness in Panslavic ideas. Such, too, was the building of the Committee —imitation baroque or something of the sort in the midst of a modern city.

The Committee was the work of a temporary, shallow, and not completely altruistic policy. However, that the reader might understand me correctly, I must add that though all of this was quite clear to me even at that time, I was far from viewing it with horror or wonderment. The fact that the Panslavic Committee was a naked instrument of the Soviet Government for influencing backward strata among the Slavs outside the Soviet Union and that its officials were dependent on and connected with both the secret and public agencies of the government—all this did not trouble me one bit. I was only disturbed by its impotence and superficiality, and above all by the fact that it could not open the way for me to the Soviet Government and to a solution of Yugoslav needs. For I too, like every other Communist, had it inculcated in me and I was convinced that there could exist no opposition between the Soviet Union and another people, especially not a revolutionary and Marxist party, as the Yugoslav Party indeed was. And though the Panslavic Committee seemed too antiquated to me, and accordingly an unsuitable instrument for a Communist end, yet I considered it acceptable, all the more so because the Soviet leadership insisted on it. As far as its officials' connections with security agencies were concerned, had I not also learned to look upon these as almost divine guardians of the revolution and of socialism —"a sword in the hands of the Party"?

The character of my insistence that I reach the summits of the Soviet Government should also be explained. Though I urged, I was neither importunate nor resentful of the Soviet Government, for I was trained to see in it something even greater than the leadership of my own Party and revolution—the leading power of Communism as a whole. I had already gathered from Tito and others that long waits—to be sure, by foreign Communists—were rather the style in Moscow. What troubled me and made me impatient was the urgency of the needs of a revolution, my own Yugoslav revolution at that.

Though nobody, not even the Yugoslav Communists, spoke of revolution, it was long since obvious that it was going on. In the West they were already writing a great deal about it. In Moscow, however, they obdurately refused to recognize it—even those who had, so to speak, every reason to do so. Everyone stubbornly talked only about the struggle against the German invaders and even more stubbornly stressed exclusively the patriotic character of that struggle, all the while conspicuously emphasizing the decisive role of the Soviet Union in the whole matter. Of course, nothing could have been further from my mind than the thought of denying the decisive role of the Soviet Party in world Communism, or of the Red Army in the war against Hitler. But on the soil of my land, and under conditions of their own, the Yugoslav Communists were obviously waging a war independent of the momentary successes and defeats of the Red Army, a war, moreover, that was at the same time converting the political and social structure of the country. Both externally and internally the Yugoslav revolution had transcended the needs and accommodations of Soviet foreign policy, and this is how I explained the obstacles and lack of understanding which I was meeting.

Strangest of all was the fact that those who should have understood this best of all submissively kept still and pretended not to understand. I had yet to learn that in Moscow the discussion and especially the determination of political positions had to wait until Stalin, or at least Molotov, had spoken. This applied even to such distinguished persons as the former secretaries of the Comintern, Manuilsky and Dimitrov.

Tito and Kardelj, as well as other Yugoslav Communists who had been to Moscow, had reported that Manuilsky was particularly well disposed toward the Yugoslavs. This may have been held against him during the purges of 1936-1937, in which almost the entire group of Yugoslav Communists had perished in the Party purge, but now, after the Yugoslav uprising against the Nazis, this could be taken for farsightedness. In any case, he injected into his enthusiasm for the Yugoslavs' struggle a certain dose of personal pride, though he knew none of the new Yugoslav leaders except, perhaps, Tito, and him only very slightly. Our meeting with him took place in the evening. Also present was G. F. Aleksandrov, the noted Soviet philosopher and, much more important, chief of the section for agitation and propaganda of the Central Committee.

Aleksandrov left no definite impression with me: Indefiniteness, or, rather, colorlessness, was his basic characteristic. He was a short, pudgy baldpate whose pallor and corpulence proclaimed that he never set foot outside his office. Except for a few conventional observations and benign smiles, he spoke not a word about the character and scope of the Yugoslav Communist uprising, though in my conversation, supposedly without design, I touched on these very points. Obviously the Central Committee had not yet determined its stand; thus, as far as Soviet propaganda was concerned, it remained simply a struggle against invaders without any real repercussions for the internal Yugoslav state or for international relations.

Nor did Manuilsky take any definite stand. Yet he exhibited a lively, emotional interest. I had already heard of his oratorical gift. One could detect this gift even in his articles, and he fairly scintillated through the polish and vividness of his expression. He was a slight and already hunched old-timer, dark-haired, with a clipped mustache. He spoke with a lisp, almost gently and—what astonished me at the time—without much energy. He was also this way in other things—considerate, affable to the point of joviality, and obviously worldly in culture.

In describing the development of the uprising in Yugoslavia, I pointed out that there was being formed in a new way a government which was in essence identical with the Soviet. I made a special point of stressing the new revolutionary role of the peasantry; I practically reduced the uprising in Yugoslavia to a tie between a peasant rebellion and the Communist avant-garde. Yet though neither he nor Aleksandrov opposed what I was saying, neither did they indicate in any way that they approved of my views. Even if I regarded it natural that Stalin's role was decisive in everything, still I expected from Manuilsky a greater independence and initiative in word and deed. I went away from my meeting with him impressed by the vitality of his personality and moved by his enthusiasm for the struggle in Yugoslavia, but also convinced that Manuilsky played no real role in the determination of Moscow's policies, not even concerning Yugoslavia.

When speaking of Stalin he attempted to camouflage extreme flattery in "scientific" and "Marxist" formulas. This manner of expression about Stalin went approximately like this: "You know, it is simply incomprehensible that a single person could have played such a decisive role in a crucial moment of the war. And that so many talents should be combined in one person—statesman, thinker, and soldier!"

My observations regarding Manuilsky's insignificance were later cruelly confirmed. He was made Foreign Minister of the Ukraine (he was a Ukrainian Jew by birth), which meant his final isolation from all substantial political activity. True, as Secretary of the Comintern he was Stalin's obedient tool, all the more because his past had not been completely Bolshevik; he had belonged to a group of so-called *mezhraiontsy,* led by Trotsky, which had joined the Bolsheviks only on the eve of the 1917 Revolution. I saw him in 1949 at the United Nations. There he came out in the name of the Ukraine against the "imperialists" and "Tito's fascist clique." Of his oratory there remained only turbulence, and of his penetrating thought only phrase-making. He was already a lost, senile little old man of whom almost every trace was lost as he slid down the steep ladder of the Soviet hierarchy.

This was not the case with Dimitrov. I met him three times during my stay—twice in the hospital of the Soviet Government, and the third time in his villa near Moscow. Each time he struck me as being a sick man. His breathing was asthmatic, the color of his skin an unhealthy red and pale, and spots around his ears were dried up as if from eczema. His hair was so sparse that it left exposed his withered yellow scalp. But his thoughts were quick and fresh, quite in contrast to his slow and tired movements. This prematurely old, almost crushed man still radiated a powerful conscious energy and vigor. His features bespoke this too, especially the strained look of his bulging bluish eyes and the convulsive protrusion of his nose and jaw. Though he did not voice his every thought, his conversation was frank and firm. It could not be said that he did not understand the situation in Yugoslavia, though he, too, regarded as premature—in view of relations between the USSR and the West—the affirmation of its factually Communist character. Of course I, too, felt that our primary propaganda effort should stress the struggle against the invader, and accordingly this meant not to accentuate the Communist char-

acter of that struggle. But it was of the utmost importance to me that the Soviet leaders, and Dimitrov too,
realize—at least regarding Yugoslavia—the senselessness of
insisting on a coalition between the Communist and bourgeois parties, inasmuch as the war and the civil war had
already shown the Communist Party to be the only real
political force. This view of mine meant nonrecognition
of the Yugoslav Royal Government-in-exile, and, in fact,
of the monarchy itself.

During our first meeting I described for Dimitrov the
developments and the situation in Yugoslavia. He gen-
erously admitted that he had not expected that the Yugoslav Party would prove to be the most militant and most
resourceful; he had placed greater hopes in the French
Party. He recalled how Tito, on leaving Moscow at the
end of 1939, swore that the Yugoslav Party would wash
away the stain with which various fractionalists had be-
smirched it and that it would prove itself worthy of the
name which it bore, whereupon Dimitrov advised him
not to swear, but to act wisely and resolutely. He recounted further: "You know, when the subject came up
of whom to appoint Secretary of the Yugoslav Party,
there was some wavering, but I was for Walter [this was
Josip Broz's Party pseudonym at the time; later he
adopted the name Tito]. He was a worker, and he seemed
solid and serious to me. I am glad that I was not mis-
taken."

Dimitrov remarked, almost apologetically, that the
Soviet Government had not been in a position to help
the Yugoslav Partisans in their greatest hour of need. He
himself had personally gotten Stalin interested in this.

That was true: as early as 1941-1942 Soviet pilots had tried to get through to Yugoslav Partisan bases, and some homeward-bound Yugoslav émigrés who had flown with them froze.

Dimitrov also mentioned our negotiations with the Germans over the exchange of prisoners: "We were afraid for you, but luckily everything turned out well."

I did not react to this, nor would I have said any more than he had confirmed, not even had he insisted on the details. But there was no danger that he would say or ask something he shouldn't; in politics all that ends well is soon forgotten.

As a matter of fact, Dimitrov did not insist on anything; the Comintern had really been dissolved, and his only job now was to gather information about Communist parties and to give advice to the Soviet Government and Party.

He told me how the idea first arose to dissolve the Comintern. It was at the time the Baltic states were annexed by the Soviet Union. It was apparent even then that the main power in the spread of Communism was the Soviet Union, and that therefore all forces had to gather directly around it. The dissolution itself had been postponed because of the international situation, to avoid giving the impression that it was being done under pressure from the Germans, with whom relations were not bad at the time.

Dimitrov was a person who enjoyed Stalin's rare regard, and, what is perhaps less important, he was the undisputed leader of the Bulgarian Communist movement. Two later meetings with Dimitrov confirmed this. At the first I described conditions in Yugoslavia to the members of the Bulgarian Central Committee, and at the second there was talk of eventual Bulgarian-Yugoslav co-operation and of the struggle in Bulgaria.

Besides Dimitrov, the meeting with the Bulgarian Central Committee was attended by Kolarov, Čhervenkov, and others. Čhervenkov had greeted me on the occasion of my first visit, though he did not remain, and I took him to be Dimitrov's private secretary. He remained in the background at this second meeting as well—silent and unobtrusive, though I was later to gain a different impression of him. I had already learned from Vlahović and others that Čhervenkov was the husband of Dimitrov's sister, that he was to have been arrested at the time of the purges—the "exposé" of the political school where he was an instructor had already been published—but he took refuge with Dimitrov. Dimitrov intervened with the NKVD and made everything in order.

The purges were especially hard on the Communist émigrés, those members of illegal parties who had no one to turn to except the Soviet. The Bulgarian émigrés were lucky that Dimitrov was Secretary of the Comintern and a person with such authority. He saved many of them. There was no one to stand behind the Yugoslavs; rather, they dug graves for one another in their race for power in the Party and in their zeal to prove their devotion to Stalin and to Leninism.

Kolarov's old age was already apparent; he was past seventy and, moreover, had been politically inactive for many years. He was a kind of relic of the violent beginnings of the Bulgarian Party. He belonged to the *"tesni"* (literally, "narrow"), the left wing of the Bulgarian Socialist Party, out of which later developed the Communist Party. In 1923 the Bulgarian Communists had given armed opposition to the military clique of General Tsankov which had just previously carried out a coup and killed the peasant leader Alexander Stambuliski. Kolarov had a massive head, more Turkish than Slavic, with chiseled features, strong nose, sensuous lips, but his thoughts were of times gone by and, I say it without rancor, of inconsequential matters. My description to Kolarov of the struggle in Yugoslavia could not be a mere analysis, but was also a horrible picture of ruins and massacres. Of some ten thousand prewar Party members, hardly two thousand were still alive, while I estimated our current losses of troops and population at around one million two hundred thousand. Yet after this recital of mine all Kolarov found it appropriate to ask me was the single question: "In your opinion, is the language spoken in Macedonia closer to Bulgarian or to Serbian?"

The Yugoslav Communist leadership had already had serious altercations with the Central Committee in Bulgaria, which held that, by virtue of the Bulgarian occupation of Yugoslav Macedonia, the organization of the Yugoslav Communist Party in Macedonia should fall to it. The dispute was finally broken off by the Comintern, which approved the Yugoslav view, but only after Germany's attack on the USSR. Nevertheless, friction over Macedonia, as well as over questions concerning the Partisan uprising against the Bulgarian occupiers, continued and got worse as the inevitable hour of the defeat of Germany, and with it of Bulgaria, approached. Vlahović, too, had observed in Moscow the pretensions of the Bulgarian

Communists regarding Yugoslav Macedonia. To tell the truth, it must be added that Dimitrov was rather different in this respect: for him the matter of prime concern was the question of Bulgarian-Yugoslav *rapprochement*. But I do not believe that even he adhered to the viewpoint that the Macedonians were a separate nationality, despite the fact that his mother was a Macedonian and that his attitude toward the Macedonians showed a marked sentimentality.

Perhaps I discharged too much bitterness when I replied to Kolarov, "I do not know whether the Macedonian language is closer to Bulgarian or Serbian, but the Macedonians are not Bulgars, nor is Macedonia Bulgarian." Dimitrov found this unpleasant. He reddened and waved his hand: "It is of no importance!" And he passed on to another question.

My memory of who attended the third meeting with Dimitrov is gone with the wind, but certainly Čhervenkov could not have been absent. The meeting took place on the eve of my return to Yugoslavia, at the beginning of June 1944. It was to be devoted to co-operation between the Yugoslav and Bulgarian Communists. But it was hardly worthwhile to discuss that, for the Bulgars in fact had no Partisan units at the time.

I insisted that military operations and the creation of Partisan units in Bulgaria should be begun, and characterized as illusions the expectation that an upheaval would take place in the Bulgarian Royal Army. I based myself on the Yugoslav experience in this: from the old Yugoslav Army the Partisans got only individual officers, while the Communist Party had to create an army of small units in the course of a very stubborn struggle. It was evident that Dimitrov, too, shared these illusions, though he did agree that the creation of Partisan units should be actively undertaken.

It was obvious that he knew something I did not know. When I stressed that even in Yugoslavia, in which the occupation had destroyed the old state apparatus, a rather long time was needed to come to terms with its remnants, he interjected, "In three or four months there will be a revolution in Bulgaria anyway; the Red Army will soon be on its borders!"

Though Bulgaria was not in a state of war with the Soviet Union, it was clear to me that Dimitrov was oriented toward the Red Army as the decisive factor. To be sure, he did not categorically declare that the Red Army would enter Bulgaria, but it was obvious that he knew even then that this would happen, and he was giving me a hint. Given Dimitrov's view and expectation, my insistence on Partisan operations and units lost any importance and meaning. The conversation came down to an exchange of opinions and brotherly greetings to Tito and the Yugoslav fighters.

It is worth recording Dimitrov's attitude toward Stalin. He, too, spoke of him with admiration and respect, but without any conspicuous flattery or reverence. His relationship to Stalin was palpably that of a revolutionary who gave disciplined submission to the leader, but a revolutionary who did his own thinking. He particularly stressed Stalin's role in the war.

He recounted: "When the Germans were outside Moscow, a general uncertainty and confusion ensued. The

Soviet Government had withdrawn to Kuibyshev. But Stalin remained in Moscow. I was with him at the time, in the Kremlin. They were taking out archives from the Kremlin. I proposed to Stalin that the Comintern direct a proclamation to the German soldiers. He agreed, though he felt no good would come of it. Soon after, I too had to leave Moscow. Stalin did not leave; he was determined to defend it. And at that most dramatic moment he held a parade in Red Square on the anniversary of the October Revolution. The divisions before him were leaving for the front. One cannot express how great a moral significance was exerted when people learned that Stalin was sitting in Moscow and when they heard his words. It restored their faith and raised their confidence, and it was worth more than a good-sized army."

On that occasion I became acquainted with Dimitrov's wife. She was a Sudeten German. This was hushed up because of the general rage against Germans to which the ordinary Russian spontaneously lent himself and which he understood more easily than antifascist propaganda.

Dimitrov's villa was tastefully luxurious. It had everything—except joy. Dimitrov's only son was dead; a portrait of the wan lad hung in the father's study. The warrior could once endure defeats and take pleasure in victories, but as an old man already at the end of his powers, Dimitrov could no longer be happy or extricate himself from the silent encircling pity that met him at every step.

4

Several months before our arrival Moscow had announced that a Yugoslav Brigade had been formed in the Soviet Union. Somewhat prior to this, Polish and then Czech units had been formed. We in Yugoslavia could not imagine how such a great number of Yugoslavs came to be in the Soviet Union when even those few political émigrés who found themselves there had largely vanished in the purges.

Now, in Moscow, everything became clear to me. The bulk of the manpower in the Yugoslav Brigade was made up of the personnel of a regiment that the Croatian quisling Pavelić had sent to the Germans at the Soviet front as a token of solidarity. But Pavelić's army had no luck there; the regiment was shattered, taken prisoner at Stalingrad, and, after the usual purification, transformed, with Commander Mesić at its head, into the Yugoslav Anti-Fascist Brigade. A few Yugoslav political émigrés were collected from hither and yon and given political posts in the Brigade, while Soviet officers—both military specialists and those from Security—took over the outfitting and checking of the men.

In the beginning the Soviet representatives insisted that the Brigade's insignia be identical with those of the Yugoslav Royal Army, but on meeting with resistance from Vlahović, they agreed to introduce the insignia of the People's Liberation Army. It was hard to agree on these insignia by way of dispatches, but Vlahović nevertheless did what he could, and we found the insignia to be the result of accommodation and compromise. On our insistence this matter too was finally settled.

There were no other essential problems concerning the Brigade except our dissatisfaction that the same commander had been kept. But the Russians defended his position by saying that he had recanted and that he had influence over his men. My impression was that Mesić was deeply demoralized and that, like many, he had simply turned his coat to save himself from a prisoner-of-war camp. He was himself dissatisfied, inasmuch as his function in the unit was conspicuously nil—purely formal.

The Brigade was located in a wood near the town of Kolomna. They lived in sod houses and drilled without regard to the cruel Russian winter. At first I was astonished at the harsh discipline that prevailed in the unit. There was a certain discrepancy, a contradiction between the aims that the unit was supposed to serve and the manner in which its men were supposed to be imbued with these aims. In our Partisan units there reigned a comradeship and solidarity, and punishment was strict only for looting and disobedience. Here everything was based on a blind submission which the Prussians of Frederick I might well have envied. However, we were not successful in changing this either, given the unyielding, harsh Soviet instructors on the one hand and, on the other, men who had only yesterday fought on the side of the Germans. We carried out an inspection, delivered a speech, discussed problems superficially, and left everything as it was, ending, to be sure, with the inevitable feast with the officers, who got drunk to a man toasting

Tito and Stalin and embracing one another in the name of Slavic brotherhood.

One of our incidental duties was to have made the first medals of the new Yugoslavia. In this we encountered complete understanding, and if the medals—especially the 1941 commemorative medal—turned out badly, it was less the fault of the Soviet factory than of our modesty and the poor quality of the sketches we had brought from Yugoslavia.

Supervision of the foreign units was carried out by NKVD General Zhukov. A slender and pale blond, still young and very resourceful, Zhukov was not without humor and a refined cynicism—not rare qualities for the members of a secret service. Concerning the Yugoslav Brigade, he told me, "It's not bad, considering the material we had to work with." And that was true. If, later in Yugoslavia, it hardly distinguished itself in engagements with the Germans, this should not be ascribed to the fighting qualities of the men as much as to the unsuitability of its organization and experience as part of an army different from the Soviet and under conditions of warfare different from those on the Eastern Front.

General Zhukov held a reception in our honor. The Military Attaché of Mexico, in conversation with me, offered aid, but unfortunately we could not figure out how it could reach our troops in Yugoslavia.

Just before my departure from Moscow, I was a guest at a dinner at General Zhukov's. He and his wife lived in a small two-room apartment. Everything was comfortable, but modest, though almost luxurious for Moscow, especially in time of war. Zhukov was an excellent civil servant and, on the basis of experience, more impressed with force than ideology as the means of realizing Communism. The relationship between us took on a certain intimacy yet at the same time a reserve, for nothing could set aside the differences in our habits and views. Political friendships are good only when each remains what he is. Before I left his apartment, Zhukov presented me with an officer's automatic gun—a modest gift, but suitable in time of war.

On the other hand, I had a quite different meeting with the organs of the Soviet Secret Service. Through Captain Kozovsky I was visited in the TsDKA by a modestly dressed little man who did not hide the fact that he was from the State Security. We arranged for a meeting on the following day, in a manner so conspiratorial that, precisely because I had been an illegal worker for so many years, I regarded it all as excessively complicated, indeed a cliché. A car awaited me in a nearby street, and after an involved ride, we transferred into another, only to be deposited in some street of the huge city from which we then walked to a third street, where someone from the window of an enormous apartment building threw down a little key which enabled us finally to enter a spacious and luxurious apartment on the third floor.

The owner of the apartment—if she was the owner—was one of those northern blonds with limpid eyes whose buxomness enhanced her beauty and strength. Her radiant beauty played no role, at least in this instance, and it turned out that she was more important than the man who brought me. She did the questioning, and he recorded the answers. They were more interested in the men who were in the councils of the Communist Party than in men of other parties. I had the uncomfortable sensation of a police interrogation, and yet I knew that it was my duty as a Communist to give the required information. Had some member of the Central Committee of the Soviet Party called me, I would not have hesitated. But what did these people want with data about the Communist Party and leading Communists when their job was to wage a struggle against the enemies of the Soviet Union and possible *provocateurs* within the Communist parties? Nevertheless I answered their questions, avoiding any precise or negative judgments, and especially any references to inner friction. I did this as much out of moral repugnance at saying things about my comrades which they would not know as out of an inner passionate aversion toward those who I felt had no right to intrude into my intimate world, my views, and my Party. My embarrassment no doubt communicated itself to my hosts, for the business part of the meeting lasted hardly an hour and a half; thereupon it turned into a less forced comradely conversation over coffee and cakes.

My contacts with the Soviet public were both more frequent and direct. At that time the public's contact with foreigners from Allied countries was not severely restricted in the Soviet Union.

Because there was a war and we were the representatives of the only Party and people who had raised a revolt against Hitler, we excited every kind of curiosity. Writers came to us for new inspiration, film producers for interesting stories, journalists for articles and information, and young men and girls who wanted our help in getting them flown to Yugoslavia as volunteers.

Pravda, their most authoritative daily, asked me for an article on the struggle in Yugoslavia, and *Novoe Vremia* one about Tito. In both cases I encountered difficulties with the editing of these articles. *Pravda* threw out almost everything that dealt with the character and political consequences of the struggle. The alteration of articles to fit the Party line was a part of our Party procedure. But it was done only when gross deviation or sensitive questions were involved. *Pravda,* however, threw out everything that had to do with the very essence of our struggle—the new regime and the social changes. It went even so far as to retouch my style, cutting out every figure of speech that was the least bit unusual, shortening sentences, and striking out turns of phrase. The article became gray and uninspired. After tussling with one of their editors, I agreed to the crippling; it was senseless to create antagonism over something like that, and it was better to publish it as it was than not at all.

The affair with *Novoe Vremia* led to even more serious trouble. Their castration of my style and my inspirations was somewhat less drastic, but they diluted or ejected practically everything that had to do with affirming the originality and extraordinary significance of Tito's personality. In my first conference with one of the editors of *Novoe Vremia,* I agreed to some immaterial changes. It was only at the second conference—when it became clear to me that in the USSR no one can be magnified except Stalin and when the editor openly admitted this in these words: "It is awkward because of Comrade Stalin; that's the way it is here"—that I agreed to the other changes; all the more so since the article had preserved its color and essence.

For me and for other Yugoslav Communists Stalin's leadership was indisputable. Yet I was nonetheless puzzled why other Communist leaders—in this case, Tito —could not be praised if they deserved it, from the Communist point of view.

It is worth noting that Tito himself was very flattered by the article and that, to the best of my knowledge, the Soviet press had never published such high praise of any other living person.

This is to be explained by the fact that Soviet public opinion—that is, the opinion of the Party, since no other kind exists—was enthusiastic about the Yugoslav struggle. But also because in the course of the war the atmosphere of Soviet society had changed.

As I look back, I can say that the conviction spread spontaneously in the USSR that now, after a war that had demonstrated the devotion of the Soviet people to their homeland and to the basic achievements of the revolution, there would be no further reason for the political restrictions and for the ideological monopolies held by little groups of leaders, and especially by a single leader. The world was changing before the very eyes of the Soviet people. It was obvious that the USSR would not be the only socialist country and that new revolutionary leaders and tribunes were making their appearance.

Such an atmosphere and such opinions did not hinder

the Soviet leaders at the time; on the contrary, these opinions contributed to the war effort. There was no reason for the leaders themselves not to encourage such illusions. After all, Tito, or, rather, the struggle of the Yugoslavs, was bringing about changes in the Balkans and in Central Europe that did not weaken the position of the Soviet Union but actually strengthened it. Thus there was no reason not to popularize and to help the Yugoslavs.

But there was an even more significant factor in this. Though allied with the Western democracies, the Soviet system, or, rather, the Soviet Communists, felt alone in the struggle. They were fighting for their own survival and exclusively for their way of life. And in view of the absence of a second front, that is, major battles in the West at a time that was decisive for the fate of the Russian people, even the ordinary man and common soldier felt alone. The Yugoslav uprising helped dispel that loneliness on the part of the leaders and the people.

Both as a Communist and as a Yugoslav I was moved by the love and regard that I encountered everywhere, especially in the Red Army. With a clear conscience I inscribed in the guest book of an exhibition of captured German weapons: "I am proud that there are no weapons here from Yugoslavia!"—for there were weapons there from all over Europe.

It was proposed that we visit the Southwestern Front —the Second Ukrainian Front—which was under the command of Marshal I. S. Konev. We went by plane to Uman, a little town in the Ukraine—and into a gashed wasteland which the war and a measureless human hatred had left in their wake.

The local Soviet arranged a supper and a meeting with the public figures of the town. The supper, which was held in a neglected, decrepit building, was hardly a gay affair. The Bishop of Uman and the Party Secretary were unable to conceal their mutual intolerance even though they were in the presence of foreigners, though both, each in his own way, were fighting against the Germans.

I had previously learned from Soviet officials that as soon as the war broke out, the Russian Patriarch began, without asking the Government, to distribute mimeographed encyclicals against the German invaders, and that they enjoyed a response which went far beyond his subordinate clergy. These appeals were also attractive in form: in the monotony of Soviet propaganda they radiated with the freshness of their ancient and religious patriotism. The Soviet Government quickly adapted itself and began to look to the Church, too, for support, despite the fact that they continued to regard it as a remnant of the old order. In the misfortunes of war, religion was revived and made headway, and the chief of the Soviet Mission in Yugoslavia, General Korneev, told how many people—and very responsible people at that—considered turning to Orthodoxy, in a moment of mortal danger from the Germans, as a more permanent ideological mobilizer. "We would have saved Russia even through Orthodoxy if that were unavoidable!" he explained.

Today this sounds incredible. But only to those who

do not comprehend the weight of the blows that smote the Russian people, to those who do not understand that every human society inevitably adopts and develops those ideas that are, at a given moment, best suited to maintaining and expanding the conditions of its existence. Though a drunkard, General Korneev was not stupid, and he was deeply devoted to the Soviet system and to Communism. To one like myself, who had grown up with the revolutionary movement and who had to fight for survival by insistence on ideological purity, Korneev's hypotheses seemed absurd. Yet I was not at all amazed— so widespread had Russian patriotism, not to say nationalism, become—when the Bishop of Uman raised a toast to Stalin as the "unifier of the Russian lands." Stalin understood intuitively that his government and his social system could not withstand the blows of the German Army unless they leaned for support on the age-old aspirations and ethos of the Russian people.

The Secretary of the Uman Soviet smoldered with bitterness at the Bishop's skillful and discreet emphasis on the role of the Church, and even more at the passive attitude of the population. The Partisan unit which he commanded was so weak in numbers that he was hardly able to deal with the pro-German Ukrainian gendarmery.

Indeed, it was not possible to conceal the passive attitude of the Ukrainians toward the war and toward Soviet victories. The population left the impression of a somber reticence, and they paid no attention to us. Although the officers with whom we were in contact covered up or embellished the behavior of the Ukrainians, our Russian chauffeur cursed their mothers because the Ukrainians had not fought better and because now the Russians had to liberate them.

The next day we set out through the Ukrainian spring mud—in the tracks of the victorious Red Army. The destroyed, twisted German equipment which we encountered so frequently added to the picture of the skill and power of the Red Army, but we marveled most of all at the toughness and self-denial of the Russian soldier, who was capable of enduring days, weeks, buried in mud up to the waist, without bread or sleep, under a hurricane of fire and steel brought by the desperate onslaughts of the Germans.

If I set aside a biased, dogmatic, and romantic enthusiasm, I would today, even as then, rate highly the qualities of the Red Army, and particularly its Russian core. True, the Soviet commanding cadres, and the soldiers and underofficers in even greater measure, receive a one-sided political education, but in every other respect they are developing initiative together with a breadth of culture. The discipline is severe and unquestioning, but not unreasonable; it is consonant with the principal aims and tasks. The Soviet officers are not only technically very proficient, but they also compose the most talented and boldest part of the Soviet intelligentsia. Though relatively well paid, they do not constitute a caste in themselves, and though not too much Marxist doctrine is required of them, they are expected all the more to be brave and not to fall back in battle—for example, the command center of the corps commander at Iaşi was three kilometers from the German lines. Stalin had carried out sweeping purges, especially in the higher commanding echelons, but these had had less effect than is sometimes believed, for he did not hesitate at the same time to elevate younger and talented men; every officer who was faithful to him and to his aims knew that his ambitions would meet with encouragement. The speed and determination with which he carried out the transformation of the top command in the midst of the war confirmed his adaptability and willingness to open careers to men of talent. He acted in two directions simultaneously: he introduced in the army absolute obedience to the Government and to the Party and to him personally, and he spared nothing to achieve military preparedness, a higher standard of living for the army, and quick promotions for the best men.

It was in the Red Army, from an army commander, that I first heard a thought that was strange to me then, but bold: When Communism triumphs in the whole world, he concluded, wars would then acquire their final bitter character. According to Marxist theories, which the Soviet commanders knew as well as I, wars are exclusively the product of class struggle, and because Communism would abolish classes, the necessity for men to wage war would also vanish. But this general, many Russian soldiers, as well as I in the worst battle in which I ever took part came to realize some further truths in the horrors of war: that human struggles would acquire the aspect of ultimate bitterness only when all men came to be subject to the same social system, for the system would be untenable as such and various sects would undertake the reckless destruction of the human race for the sake of its greater "happiness." Among these Soviet officers, trained in Marxism, this idea was incidental, tucked away. But I did not forget it, nor did I regard it as being fortuitous then. Even if their consciousness had not been penetrated by the knowledge that not even the society which they were defending was free of profound and antagonistic differences, still they vaguely discerned that though man cannot live outside an ordered society and without ordered ideas, his life is nevertheless also subject to other compelling forces.

We became inured to all sorts of things in the Soviet Union. Nevertheless, as children of the Party and the revolution who acquired faith in themselves and the faith of the people through ascetic purity, we could not help being shocked at the drinking party that was held for us on the eve of our departure from the front, in Marshal Konev's headquarters, in some village in Bessarabia.

Girls who were too pretty and too extravagantly made up to be waitresses brought in vast quantities of the choicest victuals—caviar, smoked salmon and trout, fresh cucumbers and pickled young eggplant, boiled smoked hams, cold roast pigs, hot meat pies and piquant cheeses, borsch, sizzling steaks, and finally cakes a foot thick and platters of tropical fruit under which the tables buckled.

Even earlier one could detect a concealed anticipation of the feast among the Soviet officers. Thus they all came predisposed to gorge and to guzzle. But the Yugoslavs went as if to a great trial; they had to drink, despite the fact that this was not in accord with their "Communist morality," that is, with the mores of their army and Party. Nevertheless, they comported themselves splendidly, especially considering the fact that they were not used to alcohol. A tremendous exertion of will power and conscientiousness helped them withstand many "bottoms-up" toasts, thus escaping prostration in the end.

I always drank little and cautiously, using as my excuse headaches, from which I really suffered at the time. Our General Terzić looked tragic. He had to drink even if he did not feel like it, for he did not know how to refuse a Russian confrere who would raise a toast to Stalin just a second after not having spared himself for Tito.

Our escort seemed even more tragic to me. He was a colonel from the Soviet General Staff, and because he was "from the rear," the Marshal and his generals picked on him, taking full advantage of their higher rank. Marshal Konev paid no attention to the fact that this Colonel was fairly weak; he had been brought back to work on the General Staff after having been wounded at the front. He simply commanded the Colonel: "Colonel, drink up a hundred grams of vodka to the success of the Second Ukrainian Front!" A silence ensued. All turned to the Colonel. I wanted to intercede for him. But he arose, stood at attention, and drank. Soon globules of sweat broke out on his pale high forehead.

However, not everyone drank: those who were on duty and in contact with the front did not. Nor did the staff drink at the front, except in moments of a definite lull. They said that during the Finnish campaign Zhdanov proposed to Stalin that he approve of one hundred grams of vodka a day per soldier. From that time on, the custom remained in the Red Army, except that the portion was doubled before attacks: "The soldiers feel more relaxed!" it was explained to us.

Nor did Marshal Konev drink. He had no superior to order him; besides, he had difficulties with his liver, and so his doctors forbade him to. He was a blond, tall man of fifty, with a very energetic bony face. Though he abetted gluttony, for he held to the official "philosophy" that "the men have to have a good time now and then," he himself was above that sort of thing, being sure of himself and of his troops at the front.

The author Boris Polevoy accompanied us to the front as a correspondent for *Pravda*. Though he became all too easily enthusiastic over the heroism and virtues of his country, he told us an anecdote about Konev's superhuman presence of mind and courage. Finding himself at a lookout post under fire from German mortars, Konev pretended to be looking through his binoculars, but was actually watching out of the corner of his eye to see how his officers were taking it. Every one of them knew that he would be demoted on the spot if he showed any vacillation, and no one dared point out to him the danger to his own life. And this went on. Men fell dead and wounded, but he left the post only after the inspection was over. On another occasion shrapnel struck him in the leg. They took off his boot, bandaged the leg, but he remained at the post.

Konev was one of Stalin's new wartime commanders. He was less an example of rapid promotion than Rokossovsky, for his career was neither as sudden nor as stormy as the latter's. He joined the Red Army just after

the revolution as a young worker, and gradually rose
through the ranks and through the army schools. But he,
too, made his career in battle, which was typical of the
Red Army under Stalin's leadership in the Second World
War.

Generally taciturn, Konev explained to me in a few
words the course of the campaign at Korsun'-Shevchen-
kovsky, which had just been completed and which was
compared in the Soviet Union with the one at Stalingrad.
Not without exultation, he sketched a picture of Germany's final catastrophe: refusing to surrender, some
eighty, if not even one hundred, thousand Germans were
forced into a narrow space, then tanks shattered their
heavy equipment and machine-gun nests, while the Cossack cavalry finally finished them off. "We let the Cossacks cut up as long as they wished. They even hacked
off the hands of those who raised them to surrender!"
the Marshal recounted with a smile.

I cannot say that at that moment I did not feel joy as
well over the fate that had befallen the Germans. In my
country too Nazism had, in the name of a superior race,
inflicted a war devoid of all erstwhile humane considera-
tions. And yet I had another feeling at the time—horror
that it should be so, that it could not be otherwise.

Sitting to the right of this extraordinary personality, I
was eager to clarify certain questions that interested me
in particular. First of all: Why were Voroshilov, Budënny, and other high commanders with whom the
Soviet Union entered the war shifted from their commanding positions?

Konev replied: "Voroshilov is a man of inexhaustible

courage. But—he was incapable of understanding modern warfare. His merits are enormous, but—the battle has to be won. During the Civil War, in which Voroshilov came to the fore, the Red Army had practically no planes or tanks against it, while in this war it is precisely these machines that are playing the vital role. Budënny never knew much, and he never studied anything. He showed himself to be completely incompetent and permitted awful mistakes to be made. Shaposhnikov was and remains a technical staff officer."

"And Stalin?" I asked.

Taking care not to show surprise at the question, Konev replied, after a little thought: "Stalin is universally gifted. He was brilliantly able to see the war as a whole, and this makes possible his successful direction."

He said nothing more, nothing that might sound like a stereotyped glorification of Stalin. He passed over in silence the purely military side of Stalin's direction. Konev was an old Communist firmly devoted to the Government and to the Party, but, I would say, staunch in his views on military questions.

Konev also presented us with gifts: for Tito, his personal binoculars, and for us, pistols. I kept mine until the Yugoslav authorities confiscated it at the time of my arrest in 1956.

The front abounded in examples of the personal heroism and unyielding tenacity and initiative of the common soldiers. Russia was all last-ditch resistance and deprivation and will for ultimate victory. In those days Moscow, and we with it, abandoned itself childishly to "salutes"— fireworks that greeted victories behind which loomed fire and death, and also bitterness. For this was a joy too for Yugoslav fighters suffering the misfortune of their own country. It was as though nothing else existed in the Soviet Union except this gigantic, compelling effort of a limitless land and multimillioned people. I, too, saw only them, and in my bias identified the patriotism of the Russian people with the Soviet system, for it was the latter that was my dream and my struggle.

5

It must have been about five o'clock in the afternoon, just as I had completed my lecture at the Panslavic Committee and had begun to answer questions, when someone whispered to me to finish immediately because of an important and pressing matter. Not only we Yugoslavs but also the Soviet officials had lent this lecture a more than usual importance. Molotov's assistant, A. Lozovsky, had introduced me to a select audience. Obviously the Yugoslav problem was becoming more and more acute among the Allies.

I excused myself, or they made my excuses for me, and was whisked out into the street in the middle of things. There they crammed me together with General Terzić into a strange and not very imposing car. Only after the car had driven off did an unknown colonel from the State Security inform us that we were to be received by Joseph Vissarionovich Stalin. By that time our Military Mission had been moved to a villa in Serebrennyi Bor, a Moscow suburb. Remembering the gifts for Stalin, I worried that we would be late if we went that far to get them. But the infallible State Security had taken care of that too; the gifts lay next to the Colonel in the car. Everything then was in order, even our uniforms; for some ten days or so we had been wearing new ones made in a Soviet factory. There was nothing to do but be calm and listen to the Colonel, and ask him as little as possible.

I was already accustomed to the latter. But I could not suppress my excitement. It sprang from the unfathomable depths of my being. I was aware of my own pallor and of my joyful, and at the same time almost panic-stricken, agitation.

What could be more exciting for a Communist, one who was coming from war and revolution? To be received by Stalin—this was the greatest possible recognition for the heroism and suffering of our Partisan warriors and our people. In dungeons and in the holocaust of war, and in the no less violent spiritual crises and clashes with the internal and external foes of Communism, Stalin was something more than a leader in battle. He was the incarnation of an idea, transfigured in Communist minds into pure idea, and thereby into something infallible and sinless. Stalin was the victorious battle of today and the brotherhood of man of tomorrow. I realized that it was by chance that I personally was the first Yugoslav Communist to be received by him. Still, I felt a proud joy that I would be able to tell my comrades about this encounter and say something about it to the Yugoslav fighting men as well.

Suddenly everything that had seemed unpleasant about

the USSR disappeared, and all disagreements between ourselves and the Soviet leaders lost their significance and gravity, as if they had never happened. Everything disagreeable vanished before the moving grandeur and beauty of what was happening to me. Of what account was my personal destiny before the greatness of the struggle being waged, and of what importance were our disagreements beside the obvious inevitability of the realization of our idea?

The reader should know that at that time I believed that Trotskyites, Bukharinites, and other oppositionists in the Party were indeed spies and wreckers, and that therefore the drastic measures taken against them as well as all other so-called class enemies were justified. If I had observed that those who had been in the USSR in the period of the purge in the mid-thirties tended to leave certain things unsaid, I believed these had to do with nonessentials and exaggeration: it was cutting into good flesh in order to get rid of the bad, as Dimitrov once formulated it in a conversation with Tito. Therefore I regarded all the cruelties that Stalin committed exactly as his propaganda had portrayed them—as inescapable revolutionary measures that only added to his stature and his historic role. I cannot rightly tell even today what I would have done had I known the truth about the trials and the purges. I can say with certainty that my conscience would have undergone a serious crisis, but it is not excluded that I would have continued to be a Communist—convinced of a Communism that was more ideal than the one that existed. For with Communism as an idea the essential thing is not what is being done but why. Besides it was the most rational and most intoxicating, all-embracing ideology for me and for those in my disunited and desperate land who so desired to skip over centuries of slavery and backwardness and to bypass reality itself.

I had no time to compose myself, for the car soon arrived at the gates of the Kremlin. Another officer took charge of us at this point, and the car proceeded through cold and clean courtyards in which there was nothing alive except slender budless saplings. The officer called our attention to the Tsar Cannon and Tsar Bell—those absurd symbols of Russia that were never fired or rung. To the left was the monumental bell tower of Ivan the Great, then a row of ancient cannon, and we soon found ourselves in front of the entrance to a rather low long building such as those built for offices and hospitals in the middle of the nineteenth century. Here again we were met by an officer, who conducted us inside. At the bottom of the stairs we took off our overcoats, combed ourselves in front of a mirror, and were then led into an elevator which discharged us at the second floor into a rather long red-carpeted corridor.

At every turn an officer saluted us with a loud click of the heels. They were all young, handsome, and stiff, in the blue caps of the State Security. Both now and each time later the cleanliness was astonishing, so perfect that it seemed impossible that men worked and lived here. Not a speck on the carpets or a spot on the burnished doorknobs.

Finally they led us into a somewhat small office in which General Zhukov was already waiting. A small, fat, and pock-marked old official invited us to sit down while he himself slowly rose from behind a table and went into the neighboring room.

Everything occurred with surprising speed. The official soon returned and informed us that we could go in. I thought that I would pass through two or three offices before reaching Stalin, but as soon as I opened the door and stepped across the threshold, I saw him coming out of a small adjoining room through whose open doors an enormous globe was visible. Molotov was also here. Stocky and pale and in a perfect dark blue European suit, he stood behind a long conference table.

Stalin met us in the middle of the room. I was the first to approach him and to introduce myself. Then Terzić did the same, reciting his whole title in a military tone and clicking his heels, to which our host replied—it was almost comical—by saying: "Stalin."

We also shook hands with Molotov and sat down at the table so that Molotov was to the right of Stalin, who was at the head of the table, while Terzić, General Zhukov, and I were to the left.

The room was not large, rather long, and devoid of any opulence or décor. Above a not too large desk in the corner hung a photograph of Lenin, and on the wall over the conference table, in identical carved wooden frames, were portraits of Suvorov and Kutuzov, looking very much like the chromos one sees in the provinces.

But the host was the plainest of all. Stalin was in a marshal's uniform and soft boots, without any medals except a golden star—the Order of Hero of the Soviet Union, on the left side of his breast. In his stance there was nothing artificial or posturing. This was not that majestic Stalin of the photographs or the newsreels—with the stiff, deliberate gait and posture. He was not quiet for a moment. He toyed with his pipe, which bore the white dot of the English firm Dunhill, or drew circles with a blue pencil around words indicating the main subjects for discussion, which he then crossed out with slanting lines as each part of the discussion was nearing an end, and he kept turning his head this way and that while he fidgeted in his seat.

I was also surprised at something else: he was of very small stature and ungainly build. His torso was short and narrow, while his legs and arms were too long. His left arm and shoulder seemed rather stiff. He had a quite large paunch, and his hair was sparse, though his scalp was not completely bald. His face was white, with ruddy cheeks. Later I learned that this coloration, so characteristic of those who sit long in offices, was known as the "Kremlin complexion" in high Soviet circles. His teeth were black and irregular, turned inward. Not even his mustache was thick or firm. Still the head was not a bad one; it had something of the folk, the peasantry, the paterfamilias about it—with those yellow eyes and a mixture of sternness and roguishness.

I was also surprised at his accent. One could tell that he was not a Russian. Nevertheless his Russian vocabulary was rich, and his manner of expression very vivid and plastic, and replete with Russian proverbs and sayings. As I later became convinced, Stalin was well acquainted with Russian literature—though only Russian —but the only real knowledge he had outside of Russian limits was his knowledge of political history.

One thing did not surprise me: Stalin had a sense of humor—a rough humor, self-assured, but not entirely without finesse and depth. His reactions were quick and acute—and conclusive, which did not mean that he did not hear the speaker out, but it was evident that he was no friend of long explanations. Also remarkable was his relation to Molotov. He obviously regarded the latter as a very close associate, as I later confirmed. Molotov was the only member of the Politburo whom Stalin addressed with the familiar pronoun *ty,* which is in itself significant when it is kept in mind that with Russians the polite form *vy* is normal even among very close friends.

The conversation began by Stalin asking us about our impressions of the Soviet Union. I replied: "We are enthusiastic!"—to which he rejoined: "And we are not enthusiastic, though we are doing all we can to make things better in Russia." It is engraved in my memory that Stalin used the term *Russia,* and not Soviet Union, which meant that he was not only inspiring Russian nationalism but was himself inspired by it and identified himself with it.

But I had no time to think about such things then, for Stalin passed on to relations with the Yugoslav Government-in-exile, turning to Molotov: "Couldn't we somehow trick the English into recognizing Tito, who alone is fighting the Germans?"

Molotov smiled—with a smile in which there was irony and self-satisfaction: "No, that is impossible; they are perfectly aware of developments in Yugoslavia."

I was enthusiastic about this direct, straightforward

manner, which I had not till then encountered in Soviet official circles, and particularly not in Soviet propaganda. I felt that I was at the right spot, and moreover with a man who treated realities in a familiar open way. It is hardly necessary to explain that Stalin was like this only among his own men, that is, among Communists of his line who were devoted to him.

Though Stalin did not promise to recognize the National Committee as a provisional Yugoslav government, it was evident that he was interested in its confirmation. The discussion and his stand were such that I did not even bring up the question directly; that is, it was obvious that the Soviet Government would do this immediately if it considered the conditions ripe and if developments did not take a different turn—through a temporary compromise between Britain and the USSR, and in turn between the National Committee and the Yugoslav Royal Government.

Thus this question remained unsettled. A solution had to wait and be worked out. However, Stalin made up for this by being much more positive regarding the question of extending aid to the Yugoslav forces.

When I mentioned a loan of two hundred thousand dollars, he called this a trifle, saying that we could not do much with this amount, but that the sum would be allocated to us immediately. At my remark that we would repay this as well as all shipments of arms and other equipment after the liberation, he became sincerely angry: "You insult me. You are shedding your blood, and you expect me to charge you for the weapons! I am not a merchant, we are not merchants. You are fighting for the same cause as we. We are duty bound to share with you whatever we have."

But how would the aid come?

It was decided to ask the Western Allies to establish a Soviet air base in Italy which would help the Yugoslav Partisans. "Let us try," said Stalin. "We shall see what attitude the West takes and how far they are prepared to go to help Tito."

I should note that such a base—consisting of ten transport planes, if I remember well—was soon established.

"But we cannot help you much with planes," Stalin explained further. "An army cannot be supplied by plane, and you are already an army. Ships are needed for this. And we have no ships. Our Black Sea fleet is destroyed."

General Zhukov intervened: "We have ships in the Far East. We could transfer them to our Black Sea harbor and load them with arms and whatever else is needed."

Stalin interrupted him rudely and categorically. From a restrained and almost impish person another Stalin suddenly made his appearance. "What in the world are you thinking about? Are you in your right mind? There is a war going on in the Far East. Somebody is certainly not going to miss the opportunity of sinking those ships. Indeed! The ships have to be purchased. But from whom? There is a shortage of ships just now. Turkey? The Turks don't have many ships, and they won't sell any to us anyway. Egypt? Yes, we could buy some from Egypt. Egypt will sell—Egypt would sell anything, so they'll certainly sell us ships."

Yes, that was the real Stalin, who did not mince words. But I was used to this in my own Party, and I was myself partial to this manner when it came time to reach a final decision.

General Zhukov swiftly and silently made note of Stalin's decisions. But the purchase of ships and the supplying of the Yugoslavs by way of Soviet ships never took place. The chief reason for this was, no doubt, the development of operations on the Eastern Front—the Red Army soon reached the Yugoslav border and was thus able to assist Yugoslavia by land. I maintain that at the time Stalin's intentions to help us were determined.

This was the gist of the conversation.

In passing, Stalin expressed interest in my opinion of individual Yugoslav politicians. He asked me what I thought of Milan Gavrilović, the leader of the Serbian Agrarian Party and the first Yugoslav Ambassador to Moscow. I told him: "A shrewd man."

Stalin commented, as though to himself: "Yes, there are politicians who think shrewdness is the main thing in politics—but Gavrilović impressed me as a stupid man."

I added, "He is not a politician of broad horizons, though I do not think it can be said that he is stupid."

Stalin inquired where Yugoslav King Peter II had found a wife. When I told him that he had taken a Greek princess, he shot back roguishly, "How would it be, Vyacheslav Mikhailovich, if you or I married some foreign princess? Maybe some good could come of it."

Molotov laughed, but in a restrained manner and noiselessly.

At the end I presented Stalin with our gifts. They

looked particularly primitive and wretched now. But he in no way showed any disparagement. When he saw the peasant sandals, he exclaimed: *"Lapti!"*—the Russian word for them. As for the rifle, he opened and shut it, hefted it, and remarked: "Ours is lighter."

The meeting had lasted about an hour.

It was already dusk as we were leaving the Kremlin. The officer who accompanied us obviously caught our enthusiasm. He looked at us joyously and tried to ingratiate himself with every little word. The northern lights extend to Moscow at that time of year, and everything was violet-hued and shimmering—a world of unreality more beautiful than the one in which we had been living.

Somehow that is how it felt in my soul.

6

But I was to have still another, even more significant and interesting, encounter with Stalin. I remember exactly when it occurred: on the eve of the Allied landing in Normandy.

This time too no one told me anything in advance. They simply informed me that I was to go to the Kremlin, and around nine in the evening they put me in a car and drove me there. Not even anyone in the Mission knew where I was going.

They took me to the building in which Stalin had received us, but to other rooms. There Molotov was preparing to leave. While putting on a topcoat and hat, he informed me that we were having supper at Stalin's.

Molotov is not a very talkative man. While he was with Stalin, when in a good mood, and with those who think like him, contact was easy and direct. Otherwise Molotov remained impassive, even in private conversation. Nevertheless, while in the car, he asked me what languages I spoke besides Russian. I told him that I had French. Then the conversation took up the strength and organization of the Communist Party of Yugoslavia. I emphasized that the war found the Yugoslav Party illegal and relatively few in numbers—some ten thousand members, but excellently organized. I added, "Like the Bolshevik Party in the First World War."

"You are wrong!" Molotov retorted. "The First World War found our Party in a very weak state, organizationally disconnected, scattered, and with a small membership. I remember," he continued, "how at the beginning of the war I came illegally from Petrograd to Moscow on Party business. I had nowhere to spend the night but had to risk staying with Lenin's sister!" Molotov also mentioned the name of that sister, and, if I remember correctly, she was called Maria Ilyinichna.

The car sped along at a relatively good clip—about sixty miles an hour, and met with no traffic obstacles. Apparently the traffic police recognized the car in some way and gave it a clear path. Having gotten out of Moscow, we struck out on an asphalt road which I later learned was called the Government Highway because only Government cars were permitted on it long after the war. Is this still true today? Soon we came to a barrier. The officer in the seat next to the chauffeur flashed a little badge through the windshield and the guard let us through without any formalities. The right window was down. Molotov observed my discomfort because of the draft and began to raise the window. Only then did I notice that the glass was very thick and then it occurred to me that we were riding in an armored car. I think it was a Packard, for Tito got the same kind in 1945 from the Soviet Government.

Some ten days prior to that supper the Germans had carried out a surprise attack on the Supreme Staff of the Yugoslav Army of People's Liberation in Drvar. Tito and the military missions had to flee into the hills. The Yugoslav leaders were forced to undertake long strenuous marches in which valuable time for military and political activities was lost. The problem of food also became acute. The Soviet Military Mission had been informing Moscow in detail about all this, while our Mission in Moscow was in constant contact with responsible Soviet officers, advising them how to get aid to the Yugoslav forces and the Supreme Staff. Soviet planes flew even at night and dropped ammunition and food supplies, though actually without much success, since the packages were scattered over a wide forest area which had to be quickly evacuated.

On the way Molotov wished to know my opinion regarding the situation that had arisen in connection with this. His interest was intense but without any excitement —more for the sake of obtaining a true picture.

We drove about twenty miles, turned left onto a side road, and soon came to a clump of young fir trees. Again a barrier, then a short ride, and the gate. We found ourselves before a not very large villa which was also in a thick clump of firs.

We no sooner entered a small hall from the entrance than Stalin appeared—this time in shoes and dressed in his plain tunic, buttoned up to his chin, and known so well from his prewar pictures. Like this he seemed even smaller, but also simpler and completely at home. He led us into a small and surprisingly empty study—no books, no pictures, just bare wooden walls. We seated ourselves around a small writing table, and he immediately began to inquire about events concerning the Yugoslav Supreme Staff.

The very manner of his inquiry showed a sharp contrast between Stalin and Molotov. With Molotov not only his thoughts but also the process of their generation was impenetrable. Similarly his mentality remained sealed and inscrutable. Stalin, however, was of a lively, almost restless temperament. He always questioned—himself and others; and he argued—with himself and others. I will not say that Molotov did not easily get excited, or that Stalin did not know how to restrain himself and to dissimulate; later I was to see both in these roles. But Molotov was almost always the same, with hardly a shade of variety, regardless of what or who was under consideration, whereas Stalin was completely different in his own, the Communist, milieu. Churchill has characterized Molotov as a complete modern robot. That is correct. But that is only one, external side of him. Stalin was no less a cold calculator than he. But precisely because his was a more passionate and many-sided nature—though all sides were equal and so convincing that it seemed he never dissembled but was always truly experiencing each of his roles—he was more penetrable and offered greater possibilities. The impression was gained that Molotov looked upon everything—even upon Communism and its final aims—as relative, as something to which he had to, rather than ought to, subordinate his own fate. It was as though for him there was nothing permanent, as though there was only a transitory and unideal reality which presented itself differently every day and to which he had to offer himself and his whole life. For Stalin, too, everything was transitory. But that was his philosophical view. Behind that impermanence and within it, certain great and final ideals lay hidden—his ideals, which he could approach by manipulating or kneading the reality and the living men who comprised it.

In retrospect it seems to me that these two, Molotov, with his relativism, with his knack for detailed daily routine, and Stalin, with his fanatical dogmatism and, at the same time, broader horizons, his driving quest for further, future possibilities, these two ideally complemented one another. Molotov, though impotent without Stalin's leadership, was indispensable to Stalin in many ways. Though both were unscrupulous in their methods, it seems to me that Stalin selected these methods carefully and fitted them to the circumstances, while Molotov regarded them in advance as being incidental and unimportant. I maintain that he not only incited Stalin into doing many things, but that he also sustained him and dispelled his doubts. And though, in view of his greater versatility and penetration, Stalin claims the principal role in transforming a backward Russia into a modern industrial imperial power, it would be wrong to underestimate Molotov's role, especially as the practical executive.

Molotov even seemed physically suited to such a role: thorough, deliberate, composed, and tenacious. He drank more than Stalin, but his toasts were shorter and calculated to produce a particular political effect. His personal life was also unremarkable, and when, a year later, I met his wife, a modest and gracious woman, I had the impression that any other might have served his regular, necessary functions.

The conversation with Stalin began with his excited inquiries into the further destinies of the Yugoslav Supreme Staff and the units around it. "They will die of hunger!" he exclaimed.

I tried to show him that this could not happen.

"And why not?" he went on. "How many times have soldiers been overcome by hunger! Hunger is the terrible enemy of every army."

I explained to him, "The terrain is such that something can always be found to eat. We were in much worse situations and still we were not overcome by hunger." I succeeded in calming and assuring him.

He then turned to the possibilities of sending aid. The Soviet front was still too distant to permit fighter planes to escort transports. At one point Stalin flared up, upbraiding the pilots: "They are cowards—afraid to fly during daytime! Cowards, by God, cowards!"

Molotov, who was informed on the whole problem, defended the pilots: "No, they are not cowards. Far from it. It is just that fighter planes do not have such a range and the transports would be shot down before they ever reached their target. Besides, their payload is insignificant. They have to carry their own fuel to get back. That is the only reason why they have to fly nights and carry a small load."

I supported Molotov, for I knew that Soviet pilots had volunteered to fly in daytime, without the protection of fighter planes, in order to help their fellow-soldiers in Yugoslavia.

At the same time I was in complete agreement with Stalin's insistence on Tito's need, in view of the serious and complicated circumstances and tasks, to find himself a more permanent headquarters and to free himself of daily insecurity. There is no doubt that Stalin also transmitted this view to the Soviet Mission, for it was just at that time, on their insistence, that Tito agreed to evacuate to Italy, and from there to the island of Vis, where he remained until the Red Army got to Yugoslavia. Of course Stalin said nothing about this evacuation, but the idea was taking shape in his mind.

The Allies had already approved the establishment of a Soviet air base in Italy for aid to the Yugoslav soldiers, and Stalin stressed the urgency of sending transport planes there and activating the base itself.

Apparently encouraged by my optimism regarding the final outcome of the current German offensive against Tito, he then took up our relations with the Allies, primarily with Great Britain, which constituted, as it appeared to me even then, the principal reason for the meeting with me.

The substance of his suggestions was, on the one hand, that we ought not to "frighten" the English, by which he meant that we ought to avoid anything that might alarm them into thinking that a revolution was going on in Yugoslavia or an attempt at Communist control. "What do you want with red stars on your caps? The form is not important but what is gained, and you—red stars! By God, stars aren't necessary!" Stalin exclaimed angrily.

But he did not hide the fact that his anger was not very great. It was a reproach, and I explained to him: "It is impossible to discontinue the red stars because they are already a tradition and have acquired a certain meaning among our fighters."

Standing by his opinion, but without great insistence, he turned to relations with the Western Allies from another aspect, and continued, "Perhaps you think that just because we are the allies of the English that we have forgotten who they are and who Churchill is. They find nothing sweeter than to trick their allies. During the First World War they constantly tricked the Russians and the French. And Churchill? Churchill is the kind who, if you don't watch him, will slip a kopeck out of your pocket. Yes, a kopeck out of your pocket! By God, a kopeck out of your pocket! And Roosevelt? Roosevelt is not like that. He dips in his hand only for bigger coins. But Churchill? Churchill—even for a kopeck."

He underscored several times that we ought to beware of the Intelligence Service and of English duplicity, especially with regard to Tito's life. "They were the ones who killed General Sikorski in a plane and then neatly shot down the plane—no proof, no witnesses."

In the course of the meeting Stalin kept repeating these warnings, which I transmitted to Tito upon my return and which probably played a certain role in deciding his conspiratorial night flight from Vis to Soviet-occupied territory in Rumania on September 21, 1944.

Stalin then moved on to relations with the Yugoslav Royal Government. The new royal mandatory was Dr. Ivan Šubašić, who had promised the regulation of relations with Tito and recognition of the People's Liberation Army as the chief force in the struggle against the forces of occupation. Stalin urged, "Do not refuse to hold conversations with Šubašić—on no account must you do this. Do not attack him immediately. Let us see what he wants. Talk with him. You cannot be recognized right away. A transition to this must be found. You ought to talk with Šubašić and see if you can't reach a compromise somehow."

His urging was not categorical, though determined. I transmitted this to Tito and to the members of the Central Committee, and it is probable that it played a role in the well-known Tito-Šubašić Agreement.

Stalin then invited us to supper, but in the hallway we stopped before a map of the world on which the Soviet Union was colored in red, which made it conspicuous and bigger than it would otherwise seem. Stalin waved his hand over the Soviet Union and, referring to what he had been saying just previously against the British and the Americans, he exclaimed, "They will never accept the idea that so great a space should be red, never, never!"

I noticed that on the map the area around Stalingrad was encircled from the west by a blue pencil mark. Apparently Stalin had done this in the course of the Battle of Stalingrad. He detected my glance, and I had the impression that it pleased him, though he did not betray his feelings in any way.

I do not remember the reason, but I happened to remark, "Without industrialization the Soviet Union could not have preserved itself and waged such a war."

Stalin added, "It was precisely over this that we quarreled with Trotsky and Bukharin."

And this was all—here in front of the map—that I ever heard from him about these opponents of his: they had quarreled!

In the dining room two or three people from the Soviet high echelon were already waiting, standing, though there was no one from the Politburo except Molotov. I have forgotten them. Anyway they were silent and withdrawn the whole evening.

In his memoirs Churchill vividly describes an improvised dinner with Stalin at the Kremlin. But this is the way Stalin's dinners were in general.

In a spacious and unadorned, though tasteful, dining room, the front half of a long table was covered with all kinds of foods on warmed heavy silver platters as well as beverages and plates and other utensils. Everyone served himself and sat where he wished around the free half of the table. Stalin never sat at the head, but he always sat in the same chair—the first to the left of the head of the table.

The variety of food and drink was enormous—with meats and hard liquor predominating. But everything else was simple and unostentatious. None of the servants appeared except when Stalin rang, and the only occasion for this was when I requested beer. Everyone ate what he pleased and as much as he wanted; only there was rather too much of urging and daring us to drink and there were too many toasts.

Such a dinner usually lasted six or more hours—from ten at night till four or five in the morning. One ate and drank slowly, during a rambling conversation which ranged from stories and anecdotes to the most serious political and even philosophical subjects. Unofficially and in actual fact a significant part of Soviet policy was shaped at these dinners. Besides they were the most frequent and most convenient entertainment and only luxury in Stalin's otherwise monotonous and somber life.

Apparently Stalin's co-workers were used to this manner of working and living—and spent their nights dining with Stalin or with one of their own number. They did not arrive in their offices before noon, and usually stayed in them till late evening. This complicated and made difficult the work of the higher administration, but the latter adapted itself, even the diplomatic corps, insofar as they had contacts with members of the Politburo.

There was no established order according to which members of the Politburo or other high officials attended these dinners. Usually those attended who had some connection with the business of the guest or with current issues. Apparently the circle was narrow, however, and it was an especial honor to be invited to such a dinner. Only Molotov was always present, and I maintain that this was not only because he was Commissar, that is, Minister for

Foreign Affairs, but also because he was in fact Stalin's substitute.

At these dinners the Soviet leaders were at their closest, most intimate with one another. Everyone would tell the news from his bailiwick, whom he had met that day, and what plans he was making. The sumptuous table and considerable, though not immoderate, quantities of alcohol enlivened spirits and intensified the atmosphere of cordiality and informality. An uninstructed visitor might hardly have detected any difference between Stalin and the rest. Yet it existed. His opinion was carefully noted. No one opposed him very hard. It all rather resembled a patriarchal family with a crotchety head whose foibles always caused the home folks to be apprehensive.

Stalin took quantities of food that would have been enormous even for a much larger man. He usually picked meat, which reflected his mountaineer origins. He also liked all kinds of specialties, in which this land of various climes and civilizations abounded, but I did not notice that any one food was his particular favorite. He drank moderately, most frequently mixing red wine and vodka in little glasses. I never noticed any signs of drunkenness in him, whereas I could not say the same for Molotov, and especially not for Beria, who was practically a drunkard. As all to a man overate at these dinners, the Soviet leaders ate very little and irregularly during the day, and many of them dieted on fruit and juices one day out of every week, for the sake of *razgruzhenie* (unloading).

It was at these dinners that the destiny of the vast Russian land, of the newly acquired territories, and, to a considerable degree, of the human race was decided. And

even if the dinners failed to inspire those spiritual cre-
ators—the "engineers of the human spirit"—to great
deeds, many such deeds were probably buried there for-
ever.

Still I never heard any talk of inner-Party opposition
or how to deal with it. Apparently this belonged largely
to the jurisdiction of Stalin and the Secret Police, and
since the Soviet leaders are also human, they gladly forgot
about conscience, all the more so because any appeal to
conscience would be dangerous to their own fate.

I shall mention only what seemed significant to me in
the facile and imperceptible rambling from subject to
subject at that session.

Calling to mind earlier ties between the South Slavs
and Russia, I said, "But the Russian tsars did not understand the aspirations of the South Slavs—they were interested in imperialistic expansion, and we in liberation."

Stalin agreed, but in a different way: "Yes, the Russian
tsars lacked horizons."

Stalin's interest in Yugoslavia was different from that
of the other Soviet leaders. He was not concerned with
the sacrifices and the destruction, but with what kind of
internal relations had been created and what the actual
power of the rebel movement was. He did not collect even
this information through questioning, but in the course
of the conversation itself.

At one point he expressed interest in Albania. "What
is really going on over there? What kind of people are
they?"

I explained: "In Albania pretty much the same thing
is happening as in Yugoslavia. The Albanians are the

most ancient Balkan people—older than the Slavs, and even the ancient Greeks."

"But how did their settlements get Slavic names?" Stalin asked. "Haven't they some connection with the Slavs?"

I explained this too. "The Slavs inhabited the valleys in earlier times—hence the Slavic place names—and then in Turkish times the Albanians pushed them out."

Stalin winked roguishly. "I had hoped that the Albanians were at least a little Slavic."

In telling him about the mode of warfare in Yugoslavia and its ferocity, I pointed out that we did not take German prisoners because they killed all of our prisoners.

Stalin interrupted, laughing: "One of our men was leading a large group of Germans, and on the way he killed all but one. They asked him, when he arrived at his destination: 'And where are all the others?' 'I was just carrying out the orders of the Commander in Chief,' he said, 'to kill every one to the last man—and here is the last man.' "

In the course of the conversation, he remarked about the Germans, "'They are a queer people, like sheep. I remember from my childhood: wherever the ram went, all the rest followed. I remember also when I was in Germany before the Revolution: a group of German Social Democrats came late to the Congress because they had to wait to have their tickets confirmed, or something of the sort. When would Russians ever do that? Someone has said well: 'In Germany you cannot have a revolution because you would have to step on the lawns.' "

He asked me to tell him what the Serbian words were

for certain things. Of course the great similarity between Russian and Serbian was apparent. "By God," Stalin exclaimed, "there's no doubt about it: the same people."

There were also anecdotes. Stalin liked one in particular which I told. "A Turk and a Montenegrin were talking during a rare moment of truce. The Turk wondered why the Montenegrins constantly waged war. 'For plunder,' the Montenegrin replied. 'We are poor and hope to get some booty. And what are you fighting for?' 'For honor and glory,' replied the Turk. To which the Montenegrin rejoined, 'Everyone fights for what he doesn't have.'" Stalin commented, roaring: "By God, that's deep: everyone fights for what he doesn't have."

Molotov laughed too, but again sparely and soundlessly. He was truly unable either to give or to receive humor.

Stalin inquired which leaders I had met in Moscow, and when I mentioned Dimitrov and Manuilsky, he remarked, "Dimitrov is a smarter man than Manuilsky, much smarter."

At this he remarked on the dissolution of the Comintern, "They, the Westerners, are so sly that they mentioned nothing about it to us. And we are so stubborn that had they mentioned it, we would not have dissolved it at all! The situation with the Comintern was becoming more and more abnormal. Here Vyacheslav Mikhailovich and I were racking our brains, while the Comintern was pulling in its own direction—and the discord grew worse. It is easy to work with Dimitrov, but with the others it was harder. Most important of all, there was something abnormal, something unnatural about the very existence of a general Communist forum at a time when the Communist parties should have been searching for a national language and fighting under the conditions prevailing in their own countries."

In the course of the evening two dispatches arrived; Stalin handed me both to read.

One reported what Šubašić had said to the United States State Department. Šubašić's stand was this: We Yugoslavs cannot be against the Soviet Union nor can we pursue an anti-Russian policy, for Slavic and pro-Russian traditions are very strong among us.

Stalin remarked about this, "This is Šubašić scaring the Americans. But why is he scaring them? Yes, scaring them! But why, why?"

And then he added, probably noticing the astonishment on my face, "They steal our dispatches, we steal theirs."

The second dispatch was from Churchill. He announced that the landing in France would begin on the next day. Stalin began to make fun of the dispatch. "Yes, there'll be a landing, if there is no fog. Until now there was always something that interfered. I suspect tomorrow it will be something else. Maybe they'll meet up with some Germans! What if they meet up with some Germans! Maybe there won't be a landing then, but just promises as usual."

Hemming and hawing in his customary way, Molotov began to explain: "No, this time it will really be so."

My impression was that Stalin did not seriously doubt the Allied landing, but his aim was to ridicule it, especially the reasons for its previous postponements.

As I sum up that evening today, it seems to me that I might conclude that Stalin was deliberately frightening the Yugoslav leaders in order to decrease their ties with the West, and at the same time he tried to subordinate their policy to his interests and to his relations with the Western states, primarily with Great Britain.

Thanks to both ideology and methods, personal experience and historical heritage, he regarded as sure only whatever he held in his fist, and everyone beyond the control of his police was a potential enemy. Because of the conditions of war, the Yugoslav revolution had been wrested from his control, and the power that was generating behind it was becoming too conscious of its potential for him to be able simply to give it orders. He was conscious of all this, and so he was simply doing what he could—exploiting the anticapitalist prejudices of the Yugoslav leaders against the Western states. He tried to bind those leaders to himself and to subordinate their policy to his.

The world in which the Soviet leaders lived—and that was my world too—was slowly taking on a new appearance to me: horrible unceasing struggle on all sides. Everything was being stripped bare and reduced to strife which changed only in form and in which only the stronger and the more adroit survived. Full of admiration for the Soviet leaders even before this, I now succumbed to a heady enthusiasm for the inexhaustible will and awareness which never left them for a moment.

That was a world in which there was no choice other than victory or death.

That was Stalin—the builder of a new social system.

On taking my leave, I again asked Stalin if he had any comments to make concerning the work of the Yugoslav Party. He replied, "No, I do not. You yourselves know best what is to be done."

On arriving at Vis, I reported this to Tito and to the other members of the Central Committee. And I summed up my Moscow trip: The Comintern factually no longer exists, and we Yugoslav Communists have to shift for ourselves. We have to depend primarily on our own forces.

As I was leaving after that dinner, Stalin presented me with a sword for Tito—the gift of the Supreme Soviet. To match this magnificent and exalted gift I added my own modest one, on my way back via Cairo: an ivory chess set. I do not think there was any symbolism there. But it does seem to me that even then there existed inside of me, suppressed, a world different from Stalin's.

From the clump of firs around Stalin's villa there rose the mist and the dawn. Stalin and Molotov, tired after another sleepless night, shook hands with me at the entrance. The car bore me away into the morning and to a not yet awakened Moscow, bathed in the blue haze of June and the dew. There came back to me the feeling I had had when I set foot on Russian soil: The world is not so big after all when viewed from this land. And perhaps not unconquerable—with Stalin, with the ideas that were supposed finally to have revealed to man the truth about society and about himself.

It was a beautiful dream—in the reality of war. It never

even occurred to me to determine which of these was the more real, just as I would not be able today to determine which, the dream or the reality, failed more in living up to its promises.

Men live in dreams and in realities.

II
Doubts

MY SECOND trip to Moscow, and thus my second meeting with Stalin, would probably never have taken place had I not been the victim of my own frankness.

Following the penetration of the Red Army into Yugoslavia and the liberation of Belgrade in the fall of 1944, individuals and groups within the Red Army perpetrated so many serious assaults on citizens and on members of the Yugoslav Army that a political problem arose for the new regime and for the Communist Party.

The Yugoslav Communists idealized the Red Army. Yet they themselves dealt unmercifully with even the most petty looting and crime in their own ranks. They were more dumfounded than were the ordinary people, who through inherited experience expected looting and crime by every army. The problem did in fact exist. Worse still, the foes of Communism were exploiting these incidents by Red Army soldiers in their fight against the unstabilized regime, and against Communism in general. The entire problem was complicated by the fact that the Red Army commands were deaf to complaints, and so the impression was gained that they themselves condoned the attacks and the attackers.

As soon as Tito returned to Belgrade from Rumania— at which time he also visited in Moscow and met Stalin for the first time—this question had to be taken up.

At a meeting held at Tito's, which I attended with Kardelj and Ranković—the four of us were the best-known leaders of the Yugoslav Party—it was decided to discuss this with the chief of the Soviet Mission, General Korneev. In order to have Korneev understand the whole matter in all its seriousness, it was decided that not only Tito should talk with him, but that all three of us should attend the meeting along with two of the most distinguished Yugoslav commanders—Generals Peko Dapčević and Koča Popović.

Tito presented the problem to Korneev in an extremely mild and polite form, which only made the latter's crude and offended rejection all the more astonishing. We had invited Korneev as a comrade and a Communist, and here he shouted, "In the name of the Soviet Government I protest against such insinuations against the Red Army, which has . . ."

All efforts to convince him were in vain. There suddenly loomed within him the picture of himself as the representative of a great power and of a "liberating" army.

It was then that I said, "The problem lies in the fact, too, that our enemies are using this against us and are comparing the attacks by the Red Army soldiers with the behavior of the English officers, who do not engage in such excesses."

Korneev reacted to this especially with gross lack of understanding. "I protest most sharply against the insult given to the Red Army by comparing it with the armies of capitalist countries."

Only later did the Yugoslav authorities gather statistics on the lawless acts of the Red Army soldiers. According to complaints filed by citizens, there were 121 cases of rape, of which 111 involved rape with murder, and 1,204 cases of looting with assault—figures that are hardly insignificant if it is borne in mind that the Red Army crossed only the northeastern corner of Yugoslavia. These figures show why the Yugoslav leaders had to consider these incidents as a political problem, all the more serious because it had become an issue in the domestic struggle. The Communists also regarded this problem as a moral one: Could this be that ideal and long-awaited Red Army?

The meeting with Korneev ended without results, though it was noticed later that the Soviet commands reacted more severely to the willfulness of their soldiers. As soon as Korneev left, some of the comrades reproached me, some mildly and others more sharply, for what I had said. It truly never crossed my mind to compare the Soviet Army with the British—Britain had only a mission in Belgrade—but I was stating obvious facts and presenting my reaction to a political problem, and I had been provoked too by the lack of understanding and intransigence of General Korneev. It was certainly far from my mind to insult the Red Army, which was at the time no less dear to me than to General Korneev. In view of the position I held, I could not keep silent when women were being violated—a crime I have always regarded as being among the most heinous—and when our soldiers were being abused and our property pillaged.

These words of mine, and a few other matters, were the cause of the first friction between the Yugoslav and Soviet leaders. Though actually more serious causes than these were to arise, it was precisely these words that were

to be most frequently cited as the reason for the indignation of the Soviet leaders and their representatives. I may
mention incidentally that this was certainly the reason
why the Soviet Government did not present me with the
Order of Suvorov when it distributed these to some other
leading members of the Yugoslav Central Committee.
For similar reasons it also passed over General Peko
Dapčević. This caused Ranković and me to suggest to
Tito that he decorate Dapčević with the Order of Yugoslav National Hero, to counter this snub. Those words
of mine were also one of the reasons why Soviet agents in
Yugoslavia began, in early 1945, to spread rumors about
my "Trotskyism." They themselves were forced to
abandon this measure, not just because of the senselessness of such charges, but because of an amelioration in
our relations.

Nevertheless, because of my declaration, I soon found
myself almost isolated, not particularly because my closest friends condemned me—though there were indeed
some severe reproaches—or because the Soviet leaders
had exaggerated and blown up the entire incident, but
perhaps more profoundly because of my own inner experiences. That is to say, I found myself even then in the
dilemma in which every Communist who had adopted
the Communist idea with good will and altruism finds
himself. Sooner or later he must confront the incongruity between that idea and the practice of the Party
leaders. In this case, however, it was not because of the
discrepancy between an ideal depiction of the Red Army
and the actual deeds of its members; I, too, was aware
that, though it was the army of a "classless" society, the

Red Army could "not yet" be all that it should be and that it still had to contain "remnants of the old." My dilemma was created by the indifferent, not to say benign, attitude of the Soviet leaders and Soviet commands toward crime, revealed by their refusal to recognize it and by their protests whenever it was brought to their attention. Our own intentions were good: to preserve the reputation of the Red Army and of the Soviet Union, which the propaganda of the Communist Party of Yugoslavia had been building up for years. And what did these good intentions of ours encounter? Arrogance and a rebuff typical of a big state toward a small one, of the strong toward the weak.

This dilemma was particularly reinforced and deepened by the efforts of Soviet representatives to use my basically well-intentioned words to support their arrogant critical stand toward the Yugoslav leadership.

What was it that prevented the Soviet representatives from understanding us? For what reason were my words exaggerated and twisted? Why were the Soviet representatives exploiting them in this perverted form for their political ends—to portray the Yugoslav leaders as ungrateful to a Red Army which at a given moment had allegedly played the principal role in the liberation of the capital city of Yugoslavia and had installed the Yugoslav leaders there?

But there was no answer to these questions, nor could there be at that time.

Like many others, I, too, was perturbed by other acts of the Soviet representatives. For example, the Soviet Command announced that it was presenting as aid to Belgrade a gift of a rather large quantity of wheat, but it turned out that this was in fact wheat that the Germans had collected from Yugoslav peasants and had stored on Yugoslav territory. The Soviet Command looked upon that wheat, and much else besides, simply as their spoils of war. In addition, Soviet intelligence agents recruited, en masse, émigré white Russians, and even Yugoslavs; some of these persons were right in the apparatus of the Central Committee. Against whom and why were these people employed? Also, in the field of agitation and propaganda, which I directed, friction with Soviet representatives was acutely felt. The Soviet press systematically distorted and belittled the struggle of the Yugoslav Communists, while Soviet representatives sought, at first cautiously and then more and more openly, to subordinate Yugoslav propaganda to Soviet needs and models.

And the drinking parties of the Soviet representatives, which were increasingly assuming the character of real bacchanalia and to which they were trying to entice the Yugoslav leaders, could only confirm in my eyes and in the eyes of many others the incongruity between Soviet ideals and actions, their profession of ethics in words and their amorality in deeds.

The first contact between the two revolutions and the two governments, though they were founded on similar social and ideological bases, could not but lead to friction. And since this occurred within an exclusive and closed ideology, the friction could have no other initial aspect than that of a moral dilemma, and a feeling on the part of the Yugoslavs of sorrow and regret that the center of orthodoxy did not comprehend the good intentions of a small Party and a poor land.

Inasmuch as men do not react in their consciousness, I suddenly "discovered" man's indissoluble bond with nature—I reverted to the hunting trips of my early youth and suddenly noticed that there was beauty outside of the Party and the revolution.

But the bitterness was just beginning.

2

During the winter of 1944-1945 there journeyed to Moscow a rather sizable Government delegation which included Andrija Hebrang, a member of the Central Committee and Minister of Industry, Arso Jovanović, Chief of the Supreme Staff, and Mitra Mitrović, my wife at the time. Apart from the political reactions, she was also able to relate to me the human reactions of the Soviet leaders, to which I was particularly sensitive.

The delegation, both individually and as a whole, was subjected throughout to recriminations concerning the general situation in Yugoslavia and certain of the Yugoslav leaders. The Soviet officials usually began with the correct facts, and then exaggerated them and made generalizations. To make matters worse, the chief of the delegation, Hebrang, bound himself closely to the Soviet representatives, submitting written reports to them and shifting Soviet displeasure to other members of the delegation. Hebrang was prompted to such activity, judging by everything, by his dissatisfaction at being removed from the position of Secretary of the Communist Party in Croatia, and even in greater measure because of his craven behavior while in prison—this became known only later—behavior he was trying to cover up in this manner.

To give information to the Soviet Party was at that time not in itself considered a deadly sin, for no Yugoslav Communist set his own Central Committee against the Soviet. Moreover, information on the situation in the Yugoslav Party was available and accessible to the Soviet Central Committee. However, in Hebrang's case this assumed even then the character of undermining the Yugoslav Central Committee. It was never discovered what he was reporting. But from his stand, and from what individual members of the delegation recounted, it was possible to conclude without any doubt that even at this time Hebrang was giving information to the Soviet Central Committee with the aim of getting its support and inciting it against the Yugoslav Central Committee in order to bring about changes within it that would suit him. To be sure, all of this was done in the name of principle and justified by the more or less obvious lapses and faults of the Yugoslavs. The real reason, though, lay in this: Hebrang believed that Yugoslavia should not construct its economy and economic plans independently of the USSR, while the Central Committee supported close co-operation with the USSR but not to the detriment of our own independence.

The moral *coup de grâce* to that delegation was dealt, of course, by Stalin. He assembled the entire delegation in the Kremlin and treated it to the usual feast as well as to a scene such as might be found only in Shakespeare's plays.

He criticized the Yugoslav Army and how it was administered. However, he attacked only me personally. And in what a way! He spoke agitatedly about the sufferings of the Red Army and about the horrors that it was forced to undergo fighting for thousands of kilometers through devastated country. He wept, crying out: "And such an army was insulted by no one else but Djilas! Djilas, of whom I could least have expected such a thing, a man whom I received so well! And an army which did not spare its blood for you! Does Djilas, who is himself a writer, not know what human suffering and the human heart are? Can't he understand it if a soldier who has crossed thousands of kilometers through blood and fire and death has fun with a woman or takes some trifle?"

He proposed frequent toasts, flattered one person, joked with another, needled a third, kissed my wife because she was a Serb, and again shed tears over the hardships of the Red Army and Yugoslav ingratitude.

Stalin and Molotov almost theatrically divided the roles between them according to their inclination: Molotov coldly spurred on the issue and aggravated feelings, while Stalin fell into a mood of tragical pathos. The zenith of his mood certainly came when Stalin exclaimed, kissing my wife, that he made his loving gesture at the risk of being charged with rape.

He spoke very little or not at all about Parties, Communism, Marxism, but very much about the Slavs, about the ties between the Russians and the South Slavs, and—again—about the heroic sacrifices and suffering of the Red Army.

Hearing about this, I was truly shaken and dazed. Today, it seems to me that Stalin made me the goat not so much for my "outburst," but because he intended to win me over in some way. Only my sincere enthusiasm for the Soviet Union and for himself as a personality could have prompted him in this.

Immediately upon my return to Yugoslavia I had written an article about my "Meeting with Stalin" which pleased him greatly. A Soviet representative had called my attention to the fact that in subsequent editions I ought to throw out the observation that Stalin's feet were too big and that I should stress more the intimacy between Stalin and Molotov. At the same time Stalin, who sized up people quickly and who particularly distinguished himself by his skill in exploiting people's weaknesses, must have known that he could not win me over on the basis of political ambitions, for I was indifferent to these, nor on an ideological basis, for I did not love the Soviet Party more than the Yugoslav. He could only influence me by way of my emotions—through my sincerity and my enthusiasm—and so he took that course.

But though my sensitivity and sincerity were my strong points, they easily turned into something quite opposite when I encountered insincerity and injustice. For this reason Stalin did not dare recruit me openly. I became all the more adamant and determined as experience demonstrated to me the unjust, hegemonistic Soviet intentions, that is, as I freed myself of my sentimentality.

Today it is truly difficult to ascertain how much of Stalin's action was play-acting and how much was real rancor. I personally believe that with Stalin it is impossible to separate the one from the other. With him, pretense was so spontaneous that it seemed he himself became convinced of the truth and sincerity of what he was saying. He very easily adapted himself to every turn in the discussion of any new topic, and even to every new personality.

At any rate, the delegation returned completely dazed and depressed.

Meanwhile, my isolation deepened, now also because of Stalin's tears over my "ingratitude" toward the Red Army. Though more and more isolated, I did not give in to lethargy. I turned increasingly to my pen and to books, finding within myself an escape from the difficulties and misunderstanding that beset me.

3

Time took its toll. Relations between Yugoslavia and the Soviet Union could not remain where they had been fixed by military missions and armies. Ties multiplied and relations proliferated, acquiring an increasingly defined international form.

In April a state delegation was to leave to sign a treaty of mutual assistance with the Soviet Union. The delegation was led by Tito, and he was accompanied by the Minister for Foreign Affairs, Dr. Šubašić. In the delegation there were also two economic ministers—B. Andrejev and N. Petrović. That I became part of this delegation may certainly be ascribed to the desire to liquidate the dispute over the "insult" to the Red Army by means of direct contact. Tito simply included me in the delegation, and because there were no objections from the Soviet side, I boarded the Soviet plane with the rest.

It was the beginning of April, and because of the inclement weather the plane bounced the whole time. Tito and the majority of his suite became ill. Even the pilots suffered. I, too, felt sick—but in a different way.

I felt uneasy—from the moment that I first learned of my trip up to my encounter with Stalin—as though I were a penitent of some sort. Yet I was not penitent, nor did I have any real reason for being so. Around me in Belgrade there had been created an increasingly charged atmosphere, as though I was someone who had sunk low —"made a mess of it"—and so there was nothing left for such a person but to redeem himself in some way, to throw himself solely on Stalin's generosity.

The plane neared Moscow, and the already familiar feeling of isolation welled up inside me. For the first time I felt my comrades, brothers in arms, lightly abandoning me because any contact with me might endanger their position in the Party and make it appear as though they, too, had "deviated." Even in the plane itself I was not free of this. The relationship between myself and Andrejev, made intimate by war and suffering in prison—for these reveal a man's character and human relations better than anything else—was always marked by good-natured joking and frankness. But now? He seemed to pity me, powerless to help me, while I did not dare approach him—for fear of humiliating myself, but even more for fear of forcing him into an inconvenient and unwanted fraternization with me. So, too, with Petrović, whom I knew well during my onerous life and work in the underground; our friendship was predominantly intellectual, but now I would not have dared start one of our interminable discussions of Serbian political history. As for Tito, he kept still about the whole affair, as though nothing had happened, and indicated no definite feeling or view about me. Nevertheless, I suspected that, in his own way—for political reasons—he was on my side, and that this was why he was bringing me along and why he was not taking a stand.

I was experiencing my first conflict between my simple human conscience, that is, the common human propensity for the good and the true, and the environment in which I lived and to which my daily activity bound me, namely, a movement circumscribed by its own abstract aims and fettered by its actual possibilities. This conflict did not at this time, however, take that shape in my consciousness; rather, it appeared as a clash between my good intentions to better the world and the movement to which I belonged and the lack of understanding on the part of those who made the decisions.

My anxiety grew with every moment, every yard closer to Moscow.

Beneath me sped a land whose blackness was just emerging from the melting snow, a land riven by torrents and, in many places, by bombs—desolate and uninhabited. The sky, too, was cloudy and somber, impenetrable. There was neither sky nor earth for me as I passed through an unreal, perhaps dream, world which I felt at the same time to be more real than any in which I had hitherto lived. I flew teetering between sky and earth, between conscience and experience, between desire and possibility. In my memory there has remained only that unreal and painful teetering—with not a trace of those initial Slavic feelings or even hardly any of those revolutionary raptures that marked my first encounter with the Russian, the Soviet land and its leader.

On top of everything there was Tito's airsickness. Exhausted, green, he exerted the last ounce of will power to recite his speech of greeting and to go through the ceremonies. Molotov, who headed the reception committee, shook hands with me coldly, without smiling or showing any sign of recognition. It was also unpleasant to have them take Tito to a special villa while putting the rest of us up in the Metropole Hotel.

The trials and tribulations got worse. They even assumed the proportions of a campaign.

The next day, or the day after that, the telephone in my apartment rang. A seductive female voice sounded. "This is Katia."

"Katia who?" I asked.

"It's me, Katia. Don't you remember? I have to see you. I simply must see you."

Through my head there quickly filed Katias—but I did not know one of them—and on their heels came suspicion. The Soviet Intelligence Service knew that in the Communist Party of Yugoslavia views on personal morality were strict and they were setting a trap to blackmail me later. I found it neither strange nor new that "socialist" Moscow, like every metropolis, teemed with unregistered prostitutes. I knew even better, however, that they could not make contact with high-ranking foreigners, who were tended and watched here better than anywhere on earth, unless the Intelligence Service wanted it. Apart from these thoughts, I did what I would have anyway; I said calmly and curtly, "Let me alone!"—and I put down the receiver.

I suspected that I was the only target in this transparent and smutty undertaking. Nevertheless, in view of my high rank in the Party, I felt it necessary to ascertain whether the same thing had happened to Petrović and Andrejev, and, besides, I wanted to complain to them man to man. Yes, their telephones had rung too, but instead of a Katia, it was a Natasha and a Vova! I explained my own experience, and practically ordered them not to make any contact.

I had mixed feelings—relief that I was not the only target, but also deepening doubts. Why all of this? It never occurred to me to inquire of Dr. Šubašić whether a similar attempt had been directed at him. He was not a Communist, and it would be awkward for me to display the Soviet Union and its methods in a bad light before him, all the more so since they were aimed against Communists. I was quite certain, though, that no Katia had approached Šubašić.

I was not yet able to draw the conclusion—that it was precisely the Communists who were the butt and the means by which Soviet hegemony was to ensconce itself in the countries of Eastern Europe. Yet I suspected as much. I was horrified by such methods and resented having my character subjected to such manipulation.

At that time I was still capable of believing that I could be a Communist and remain a free man.

4

Nothing significant occurred in connection with the treaty of alliance between Yugoslavia and the USSR. The treaty was the usual thing, and my job was simply to verify the translation.

The signing took place in the Kremlin on the evening of April 11, in a very narrow official circle. Of the public —if such an expression may be applied to that environment—only Soviet cameramen were in attendance.

The sole striking episode occurred when Stalin, holding a glass of champagne, turned to a waiter and invited him to clink glasses. The waiter became embarrassed, but when Stalin uttered the words: "What, you won't drink to Soviet-Yugoslav friendship?" he obediently took the glass and drank it bottoms up. There was something demagogic, even grotesque, about the entire scene, but everyone looked upon it with beatific smiles, as an expression of Stalin's regard for the common people and his closeness to them.

This was my first opportunity to meet Stalin again. His mien was ungracious, though it did not have Molotov's frigid stiffness and artificial amiability. Stalin did not address a single word to me personally. The dispute over the behavior of the Red Army soldiers was obviously neither forgotten nor forgiven. I was left to go on twirling over the fire of purgatory.

Nor did he say anything at the dinner for the inner circle, in the Kremlin. After dinner we looked at movies. Because of Stalin's remark that he was tired of gunfire, they put on, not a war film, but a shallow happy collective-farm movie. Throughout the showing Stalin made comments—reactions to what was going on, in the manner of uneducated men who mistake artistic reality for actuality. The second film was a prewar one on a war theme: "If War Comes Tomorrow" (*"Esli zavtra voina . . ."*). The war in that film was waged with the help of poisonous gas, while at the rear of the invaders— the Germans—rebellious elements of the proletariat were breaking out. At the end of the film Stalin calmly remarked, "Not much different from what actually happened, only there was no poisonous gas and the German proletariat did not rebel."

Everyone was tired of toasts, of food, of films. Again without a word, Stalin shook hands with me too, but by now I was more nonchalant and calm, though I could not say why. Perhaps because of the easier atmosphere. Or was it my own inner determination and resolution? Probably both. In any event—life is possible without Stalin's love.

A day or two later there was a formal dinner in Catherine Hall. According to Soviet protocol at the time, Tito was seated to the left of Stalin and to the right of Kalinin, then President of the Supreme Soviet. I was seated at Kalinin's left. Molotov and Šubašić sat opposite

Stalin and Tito, while the other Yugoslav and Soviet officers sat around in a circle.

The stiff atmosphere seemed all the more unnatural because all present, except Dr. Šubašić, were Communists, yet they addressed one another as "Mister" in their toasts and adhered strictly to international protocol, as though this was a meeting of the representatives of differing systems and ideologies.

Aside from the toasts and the protocol, we acted like comrades toward one another, that is, like men who were close to one another, men who were in the same movement, with the same aims. This contrast between formality and reality was all the more drastic because relations between the Soviet and Yugoslav Communists were still cordial, unmarred by Soviet hegemonism and competition for prestige in the Communist world. However, life is no respecter of desires or designs, but imposes patterns which no one is capable of foreseeing.

Relations between the Soviet Union and the Western Allies were still in their wartime honeymoon, and the Soviet Government wished, by observing this formality, to avoid complaints that they were not treating Yugoslavia as an independent nation just because it was Communist. Later, after it had become entrenched in Eastern Europe, the Soviet Government was to insist on dropping protocol and other formalities as "bourgeois" and "nationalist" prejudices.

Stalin broke the ice. Only he could do it, for only he was not exposed to the danger of being criticized for a *faux pas*. He simply stood, lifted his glass, and addressed Tito as "Comrade," adding that he would not call him

"Mister." This restored real amity and livened up the atmosphere. Dr. Šubašić, too, smiled happily, though it was difficult to believe that he was doing so sincerely; pretense was not lacking in this politician, who was without ideas and without any stable foundations whatever.

Stalin began to make jokes, to direct sallies and thrusts across the table, and to grumble cheerfully. Once revived, the atmosphere did not flag.

Old Uncle Kalinin, who could barely see, had difficulty finding his glass, plate, bread, and I kept helping him solicitously the whole time. Tito had paid him a protocol visit just an hour or two before and had told me that the old man was not entirely senile. But from what Tito had reported, and from the remarks Kalinin made at the banquet, one could conclude the opposite.

Stalin certainly knew of Kalinin's decrepitude, for he made heavy-footed fun of him when the latter asked Tito for a Yugoslav cigarette. "Don't take any—those are capitalist cigarettes," said Stalin, and Kalinin confusedly dropped the cigarette from his trembling fingers, whereupon Stalin laughed and his physiognomy took on the expression of satyr. A bit later none other than Stalin raised a toast in honor of "our President," Kalinin, but these were polite phrases obviously picked for someone who for long had been nothing more than a mere figurehead.

Here, in a rather broader and more official circle, the deification of Stalin was more palpable and direct. Today I am able to conclude that the deification of Stalin, or the "cult of the personality," as it is now called, was at least as much the work of Stalin's circle and the bureaucracy, who required such a leader, as it was his own doing. Of course, the relationship changed. Turned into a deity, Stalin became so powerful that in time he ceased to pay attention to the changing needs and desires of those who had exalted him.

An ungainly dwarf of a man passed through gilded and marbled imperial halls, and a path opened before him, radiant, admiring glances followed him, while the ears of courtiers strained to catch his every word. And he, sure of himself and his works, obviously paid no attention to all this. His country was in ruins, hungry, exhausted. But his armies and marshals, heavy with fat and medals and drunk with vodka and victory, had already trampled half of Europe under foot, and he was convinced they would trample over the other half in the next round. He knew that he was one of the cruelest, most despotic personalities in human history. But this did not worry him one bit, for he was convinced that he was executing the judgment of history. His conscience was troubled by nothing, despite the millions who had been destroyed in his name and by his order, despite the thousands of his closest collaborators whom he had murdered as traitors because they doubted that he was leading the country and people into happiness, equality, and liberty. The struggle had been risky, long, and all the more underhanded because the opponents were few in number and weak. But he succeeded, and success is the only criterion of truth! For what is conscience? Does it even exist? It had no place in his philosophy, much less in his actions. After all, man is the product of productive forces.

Poets were inspired by him, orchestras blared cantatas

in his honor, philosophers in institutes wrote tomes about his sayings, and martyrs died on scaffolds crying out his name. Now he was the victor in the greatest war of his nation and in history. His power, absolute over a sixth of the globe, was spreading farther without surcease. This convinced him that his society contained no contradictions and that it exhibited superiority to other societies in every way.

He joked, too, with his courtiers—"comrades." But he did not do this exclusively out of a ruler's generosity. Royal generosity was visible only in the manner in which he did this: his jokes were never at his own expense. No, he joked because he liked to descend from his Olympian heights; after all, he lived among men and had to show from time to time that the individual was nothing without the collective.

I, too, was swept up by Stalin and his witticisms. But in one little corner of my mind and of my moral being I was awake and troubled: I noticed the tawdriness, too, and could not accept inwardly Stalin's manner of joking— nor his deliberate avoidance of saying a single human, comradely word to me.

5

Still I was pleasantly surprised when I, too, was taken to an intimate dinner in Stalin's villa. Of course Dr. Šubašić knew absolutely nothing about it. Only we Yugo-slav Communist ministers were there, and, on the Soviet side, Stalin's closest associates: Malenkov, Bulganin, General Antonov, Beria, and, to be sure, Molotov.

As usual, at about ten o'clock at night we found ourselves around Stalin's table. I had arrived in the car with Tito. At the head of the table sat Beria, to his right Malenkov, then I and Molotov, then Andrejev and Petrović, while to the left sat Stalin, Tito, Bulganin, and General Antonov, Assistant Chief of the General Staff.

Beria was also a rather short man—in Stalin's Politburo there was hardly anyone taller than himself. He, too, was somewhat plump, greenish pale, and with soft damp hands. With his square-cut mouth and bulging eyes behind his pince-nez, he suddenly reminded me of Vujković, one of the chiefs of the Belgrade Royal Police who specialized in torturing Communists. It took an effort to dispel the unpleasant comparison, which was all the more nagging because the similarity extended even to his expression—that of a certain self-satisfaction and irony mingled with a clerk's obsequiousness and solicitude. Beria was a Georgian, like Stalin, but one could not tell this at all by the looks of him. Georgians are generally bony and dark. Even in this respect he was nondescript. He could have passed more easily for a Slav or a Lett, but mostly for a mixture of some sort.

Malenkov was even smaller and plumper, but a typical Russian with a Mongol admixture—dark, with prominent cheekbones, and slightly pock-marked. He gave one the impression of being a withdrawn, cautious, and not very personable man. It seemed as though under the layers and rolls of fat there moved about still another man, lively and adept, with intelligent and alert black eyes.

He had been known for some time as Stalin's unofficial stand-in in Party matters. Practically all matters pertaining to Party organization and the promotion and demotion of officials were in his hands. He was the one who had invented "cadre lists"—detailed biographies and autobiographies of all members and candidates of a Party of many millions—which were guarded and systematically maintained in Moscow. I took advantage of my meeting with him to ask for Stalin's work *On the Opposition* (*Ob oppozitsii*), which had been withdrawn from public circulation because of the numerous citations from Trotsky, Bukharin, and others it contained. The next day I received a used copy of the work, and it is now in my library.

Bulganin was in a general's uniform. He was rather stout, handsome, and unmistakably Russian, with an old-fashioned goatee, and extremely reserved in his expression. General Antonov was still young, very handsome, dark and lithe. He, too, did not mix into the conversation unless it concerned him.

Seated across from Stalin, face to face, I suddenly gained confidence, though he did not turn to me for a long time. Not until the atmosphere had been warmed by liquor, toasts, and jesting did Stalin find the time ripe to liquidate the dispute with me. He did it in a half-joking manner. He filled for me a little glass of vodka and bade me drink to the Red Army. Not understanding his intention immediately, I wished to drink to his health. "No, no," he insisted, while smiling and regarding me probingly, "but just for the Red Army! What, you won't drink to the Red Army?"

I drank, of course, though even at Stalin's I avoided drinking anything but beer, first, because alcohol did not agree with me, and, second, because drunkenness did not agree with my views, though I was never a proponent of temperance.

Thereupon Stalin asked me about the affair of the Red Army. I explained to him that it had not been my intention to insult the Red Army, but that I had wished to call attention to irregularities of certain of its members and to the political difficulties they were creating for us.

Stalin interrupted: "Yes, you have, of course, read Dostoevsky? Do you see what a complicated thing is man's soul, man's psyche? Well then, imagine a man who has fought from Stalingrad to Belgrade—over thousands of kilometers of his own devastated land, across the dead bodies of his comrades and dearest ones! How can such a man react normally? And what is so awful in his having fun with a woman, after such horrors? You have imagined the Red Army to be ideal. And it is not ideal, nor can it be, even if it did not contain a certain percentage of criminals—we opened up our penitentiaries and stuck everybody into the army. There was an interesting case. An Air Force major had fun with a woman, and a chivalrous engineer appeared to protect her. The Major drew a gun: 'Ekh, you mole from the rear!'—and he killed the chivalrous engineer. They sentenced the Major to death. But somehow the matter was brought before me, and I made inquiries—I have the right as commander in chief in time of war—and I released the Major and sent him to the front. Now he is one of our heroes. One has to understand the soldier. The Red Army is not ideal. The important thing is that it fights Germans—and it is fighting them well, while the rest doesn't matter."

Soon after, upon my return from Moscow, I heard, to my horror, of a far more significant example of Stalin's "understanding" attitude toward the sins of Red Army personnel. Namely, while crossing East Prussia, Soviet soldiers, especially the tank units, pounded and regularly killed all German civilian refugees—women and children. Stalin was informed of this and asked what should be done. He replied: "We lecture our soldiers too much; let them have some initiative!"

That night at his villa, he then asked: "And what about General Korneev, the chief of our Mission, what kind of man is he?"

I avoided saying anything bad about him and about his Mission, though all sorts of things could have been brought up, but Stalin himself concluded: "The poor man is not stupid, but he is a drunkard, an incurable drunkard!"

After that Stalin even joked with me, on seeing that I was drinking beer. As a matter of fact, I don't even like beer. Stalin commented: "Djilas here drinks beer like a German, like a German—he is a German, by God, a German."

I did not find this joke at all to my liking; at that time hatred for the Germans, even for those few Communist émigrés, was at its height in Moscow, but I took it without anger or inner resentment.

With this, it appeared, the dispute over the behavior of the Red Army was resolved. Stalin's relation to me got back on the original track of cordiality.

And so it went, until the rift between the Yugoslav and Soviet Central Committees, in 1948, when Molotov and Stalin dredged up in their letters that same dispute over the Red Army and the "insults" that I had dealt it.

Stalin teased Tito with obvious deliberateness—in a manner that had in it as much malice as jest. He did it by speaking unfavorably of the Yugoslav Army while flattering the Bulgarian Army. That previous winter Yugoslav units including many draftees who were engaged for the first time in very serious frontal attacks had suffered defeats, and Stalin, who was apparently well informed, took the opportunity to point out, "The Bulgarian Army is better than the Yugoslav. The Bulgars had their weaknesses and enemies in their ranks. But they executed a few score—and now everything is in order. The Bulgarian Army is very good—drilled and disciplined. And yours, the Yugoslav—they are still Partisans, unfit for serious front-line fighting. Last winter one German regiment broke up a whole division of yours. A regiment beat a division!"

A bit later Stalin proposed a toast to the Yugoslav Army, but he did not forget to add to it, "But which will yet fight well on level ground!"

Tito had kept from reacting to Stalin's comments. Whenever Stalin made some witty remark at our expense, Tito looked at me silently with a restrained smile, and I returned his look with solidarity and sympathy. But when Stalin said that the Bulgarian Army was better than the Yugoslav, Tito could not stand it, and shouted that the Yugoslav Army would quickly rid itself of its weaknesses.

One could detect in the relation between Stalin and Tito something special, tacit—as though these two had a grudge against one other, but each was holding back for his own reasons. Stalin took care not to offend Tito personally in any way, but at the same time he kept taking sideswiping jabs at conditions in Yugoslavia. On the other hand, Tito treated Stalin with respect, as one would one's senior, but resentment could also be detected, especially at Stalin's remarks on Yugoslav conditions.

At one point Tito brought out that there were new phenomena in socialism and that socialism was now being achieved in ways different from those of the past, which gave Stalin an opportunity to say, "Today socialism is possible even under the English monarchy. Revolution is no longer necessary everywhere. Just recently a delegation of British Labourites was here, and we talked about this in particular. Yes, there is much that is new. Yes, socialism is possible even under an English king."

As is known, Stalin never upheld such a view publicly. The British Labourites soon gained a majority at the elections and nationalized over twenty per cent of the industrial production. Nevertheless, Stalin never recognized these measures as being socialistic nor the Labourites as being socialists. I maintain that he did not do so primarily because of differences and clashes with the Labour Government in foreign policy.

In the course of the conversation about this, I interjected that in Yugoslavia there existed in essence a Soviet type of government; the Communist Party held all the key positions and there was no serious opposition party.

But Stalin did not agree with this. "No, your government is not Soviet—you have something in between De Gaulle's France and the Soviet Union."

Tito remarked that in Yugoslavia something new was taking shape. But this discussion remained unfinished. Within myself I could not agree with Stalin's view; neither did I think that I differed with Tito.

Stalin presented his views on the distinctive nature of the war that was being waged: "This war is not as in the past; whoever occupies a territory also imposes on it his own social system. Everyone imposes his own system as far as his army can reach. It cannot be otherwise."

He also pointed out, without going into long explanations, the meaning of his Panslavic policy. "If the Slavs keep united and maintain solidarity, no one in the future will be able to move a finger. Not even a finger!" he repeated, emphasizing his thought by cleaving the air with his forefinger.

Someone expressed doubt that the Germans would be able to recuperate within fifty years. But Stalin was of a different opinion. "No, they will recover, and very quickly. That is a highly developed industrial country with an extremely qualified and numerous working class and technical intelligentsia. Give them twelve to fifteen years and they'll be on their feet again. And this is why the unity of the Slavs is important. But even apart from this, if the unity of the Slavs exists, no one will dare move a finger."

At one point he got up, hitched up his pants as though he was about to wrestle or to box, and cried out almost in a transport, "The war shall soon be over. We shall recover in fifteen or twenty years, and then we'll have another go at it."

There was something terrible in his words: a horrible war was still going on. Yet there was something impressive, too, about his cognizance of the paths he had to take, the inevitability that faced the world in which he lived and the movement that he headed.

The rest of what was said that evening was hardly worth remembering. There was much eating, even more drinking, and countless senseless toasts were raised.

Molotov recounted how Stalin stung Churchill. "Stalin raised a toast to secret agents and to the Secret Service, thus alluding to Churchill's failures at Gallipoli in the First World War, which occurred because the British lacked sufficient information." Molotov also cited, not without glee, Churchill's bizarre sense of humor. "Churchill had declared in Moscow, in his cups, that he deserved the highest order and citation of the Red Army because he had taught it to fight so well, thanks to the intervention at Archangel." One could tell in general that Churchill had left a deep impression on the Soviet leaders as a farsighted and dangerous "bourgeois statesman"—though they did not like him.

During the ride back to his villa, Tito, who also could not stand large quantities of liquor, remarked in the automobile: "I don't know what the devil is wrong with these Russians that they drink so much—plain decadence!" I, of course, agreed with him and tried in vain, who knows after how many attempts, to find an explanation of why Soviet high society drank so desperately and determinedly.

On returning to town from the villa in which Tito was housed, I collected my impressions of that night in which actually nothing significant had happened: there were no points of disagreement, and yet we seemed farther apart than we ever were. Every dispute had been resolved for political reasons, as something hardly to be avoided in relations between independent states.

At the end of our visit (following the dinner with Stalin), we spent an evening at Dimitrov's. To fill it up with something, he invited two or three Soviet actors, who gave short performances.

Of course there was talk of a future union between Bulgaria and Yugoslavia, but it was very general and brief. Tito and Dimitrov exchanged Comintern reminiscences. All in all, it was more a friendly gathering than a political meeting.

Dimitrov was alone at the time because all the Bulgarian émigrés had long since gone to Bulgaria—in the footsteps of the Red Army. One could tell that Dimitrov was tired and listless, and we knew at least part of the reason, though nothing was said about it. Although Bulgaria had been liberated, Stalin would not permit Dimitrov's return, with the excuse that it was not yet the right time, for the Western states would take his return as an open sign of the establishment of Communism in Bulgaria—as though such a sign was not evident even without this! There had been talk of this, too, at Stalin's dinner. Winking noncommittally, Stalin had said, "It is not yet time for Dimitrov to go to Bulgaria: he's well off where he is."

Though there was nothing to prove it, still it was suspected even then that Stalin was preventing Dimitrov's return until he himself settled affairs in Bulgaria! These suspicions of ours did not yet imply Soviet hegemony, though there were premonitions of this too, but we saw the matter as a necessary accommodation with Stalin's alleged fears that Dimitrov might push matters toward the left too soon in Bulgaria.

But even this was significant and sufficient—for a beginning. It evoked a whole series of questions. Stalin was a genius, but Dimitrov was hardly a nobody. By what token did Stalin know better than Dimitrov what ought to be done in Bulgaria? Did not holding Dimitrov in Moscow against his will undermine his reputation among Bulgarian Communists and the Bulgarian people? And, in general, why this intricate game over his return, in which the Russians were not accountable to anyone, not even to Dimitrov?

In politics, more than in anything else, the beginning of everything lies in moral indignation and in doubt of the good intentions of others.

6

We returned via Kiev, and at our wish and that of the Soviet Government we remained two or three days to visit the Ukrainian Government.

The Secretary of the Ukrainian Party and Premier of the Government was N. S. Khrushchev, and his Commissar for Foreign Affairs was Manuilsky. It was they who met us and it was with them that we spent the entire three days.

At the time, in 1945, the war was still on and one was permitted to express modest wishes. Khrushchev and Manuilsky expressed one—that the Ukraine might establish diplomatic relations with the "people's democracies."

However, nothing came of it. Stalin soon enough encountered resistance even in the "people's democracies," so that it hardly would occur to him to strengthen any Ukrainian separateness. As for the eloquent lively old veteran Manuilsky—a minister without a ministry—he later gave speeches in the United Nations for two or three years, only to disappear one day and to sink into the anonymous mass of the victims of Stalin's or someone else's displeasure.

Khrushchev's destiny was quite different. But at that moment no one could have surmised it. Even then he was in the top political leadership—and had been since 1939—though it was considered that he was not as close to Stalin as Molotov and Malenkov were, or even Kaganovich. In Soviet top echelons he was held to be a very skillful operator with a great capacity for economic and organizational matters, though not as a writer or speaker. He came to leadership in the Ukraine after the purges of the mid-thirties, but I am not acquainted with—nor was I then interested in—his part in them. But it is known how one rose in Stalin's Russia: certainly by dint of determination and dexterity during the bloody "anti-kulak" and "anti-Party" campaigns. This would have had to be especially true for the Ukraine, where in addition

to the afore-mentioned "deadly sins" there was "nation·
alism" as well.

Though he had achieved success while still relatively
young, there was nothing surprising about Khrushchev's
career in the light of Soviet conditions: he made his way
through schools, political and other, as a worker, and
climbed the Party ladder by means of his devotion, alert-
ness, and intelligence. Like most of the leaders, he belonged to the new postrevolutionary Stalinist generation of Party and Soviet officials. The war found him in
the highest position in the Ukraine. Because the Red
Army had to withdraw from the Ukraine before the
Germans, he was given a high political post in it, but not
the highest—he was still in the uniform of a lieutenant
general. He returned as chief of the Party and the Government in Kiev after the expulsion of the Germans.

We had heard somewhere that he was not a Ukrainian
by birth, but a Russian. Though nothing was said about
this, he himself avoided mentioning it, for it would have
been embarrassing if not even the Premier of the Ukrainian Government was a Ukrainian! It was indeed unusual
even for us Communists, who were able to justify and
explain away everything that might becloud the ideal
picture of ourselves, that among the Ukrainians, a nation
as numerous as the French and in some ways more cultured than the Russian, there was not a single person
capable of being premier of the Government.

Nor could it be concealed from us that the Ukrainians
had deserted en masse from the Red Army as the Germans
advanced into their regions. After the expulsion of the
Germans, some two and a half million Ukrainians were

drafted into the Red Army. Although minor operations were still being carried out against Ukrainian nationalists (one of their victims was the gifted Soviet General Vatutin), we still could not quite accept the explanation that this state of affairs in the Ukraine was caused exclusively by the stubbornness of Ukrainian bourgeois nationalism. The question imposed itself: Whence this nationalism if the peoples of the USSR are really equal?

We were bewildered and astonished at the marked Russification of public life. Russian was spoken in the theater, and there were even daily newspapers in Russian.

However, it was far from our intention to blame our solicitous host, N. S. Khrushchev, for this or anything else, for, as a good Communist, he could do nothing else but carry out the orders of his Party, his Leninist Central Committee, and his leader and teacher, J. V. Stalin. All Soviet leaders have distinguished themselves by their practicality and, in Communist circles, by their directness. N. S. Khrushchev stood out from the rest in both respects.

Neither then nor now—after carefully reading his speeches at congresses—did I have the impression that his knowledge went beyond the limits of classical Russian literature and Russian history, while his grasp of theory was on the level of an intermediate Party school. Beside this external knowledge gathered from courses, much more important is the knowledge that he gained as an autodidact, by constantly improving himself, and, even more, the experience he gained from his lively and many-sided activities. It is impossible to determine the quantity and quality of that knowledge, for equally astonishing is his knowledge of some rare fact and his ignorance of some elementary truths. His memory is excellent, and he expresses himself vividly and graphically.

Unlike other Soviet leaders, he exhibited an unrestrained garrulity, although like them he was fond of using folk proverbs and sayings. This was a kind of fashion at the time and proof of one's ties with the people. With him, however, there was less artificiality about this because of his naturally simple and unaffected behavior and manner of speaking. He also had a sense of humor. Unlike Stalin's humor, which was predominantly intellectual and, as such, cynical, Khrushchev's humor was typically folksy and thus often almost crude, but it was lively and inexhaustible. Now that he has attained the most exalted heights of power and is in the gaze of the whole world, one can tell that he is careful of his pose and manner of expression, but he has remained basically unchanged. Beneath the present Soviet chief of state and Party it is not difficult to discern a man of the popular masses. Yet it should be added that he suffers less than any Communist autodidact and unfinished scholar from a feeling of inferiority, that is, he does not feel the need to hide his personal ignorance and weaknesses behind an external brilliance and generalizations. The commonplaces with which his conversation abounds are the expression of both real ignorance and Marxist maxims learned by rote, but even these he presented with conviction and frankness. The language and manner with which he expresses himself encompass a broader circle than the one to which Stalin spoke, though he, too, addressed himself to the same—Party—public.

In his not very new, unpressed general's uniform, he was the only one among the Soviet leaders who delved into details, into the daily life of the Communist rank and file and the citizenry. Let it be understood: he did not do this with the aim of changing conditions, but of strengthening, perfecting existing conditions. He did look into matters and remedy them, while others issued orders from offices and received reports.

None of the Soviet leaders went to collective farms, except occasionally to attend some feast or parade. Khrushchev accompanied us to a collective farm and, without harboring in any little corner of his mind the slightest doubt of the justice of the system itself, he not only clinked huge glasses of vodka with the collective farmers, but he also inspected the garden hotbeds, peeked into the pigsty, and began discussing practical problems. During the ride back to Kiev he kept coming back to the question of the collective farms and openly brought out shortcomings.

We could observe his extraordinarily practical sense on a grand scale at a meeting of the economic sections of the Ukrainian Government. Unlike Yugoslav ministers, his commissars were excellently acquainted with matters and, what was more important, they realistically gauged possibilities.

Rather short and stocky, but brisk and agile, he was strongly hewn and of one piece. He practically bolted down impressive quantities of food—as though wishing to spare his artificial steel jawbone. While Stalin and his entourage gave the impression of gourmandism, it seemed to me that it was all the same to Khrushchev what he ate and that the important thing was to fill up, as it is to any hard worker, if, of course, he has the means. His board was also opulent—stately but impersonal. Khrushchev is not a gourmand, though he eats no less than Stalin and drinks even more.

He possesses an extremely powerful vitality and, like all practical men, a great ability to adapt. I do not think he would trouble himself much over the choice of methods as long as they brought him practical results. But like all popular demagogues who often themselves believe what they say, he would find it easy to abandon impractical methods and readily justify the change by appeals to moral reasons and the highest ideals. He likes to quote the proverb "In a fight don't stop to pick cudgels." It serves him well to justify the cudgel even when there is no fight. Everything I have said here is not at all what one should tell about Khrushchev today. Still I have given my impressions from another time, and also, along the way, my incidental reflections of today.

At that time I could not detect in Khrushchev any disapproval of Stalin or Molotov. Whenever there was talk of Stalin, he spoke of him with respect and stressed their closeness. He recounted how, on the eve of the German attack, Stalin had phoned him from Moscow warning him to be on the alert, for he had information that the Germans might begin operations the next day—June 22. I offer this as a fact, and not in order to refute Khrushchev's charges against Stalin concerning the unexpectedness of the German attack. That unexpectedness was the consequence of Stalin's error in political judgment.

Nevertheless, in Kiev one felt a certain freshness—

thanks to Khrushchev's limitless vigor and practicality, to Manuilsky's enthusiasm, to the beauty of the city itself, which, with its unobstructed horizons and with its hills overlooking a vast muddy river, was reminiscent of Belgrade. Though Khrushchev left the impression of strength, self-confidence, and realism, and Kiev one of conscious and cultivated beauty, the Ukraine has remained associated in my memory with a loss of personality, with weariness and hopelessness.

The more I delved into the Soviet reality, the more my doubts multiplied. The reconciliation of that reality and my, human, conscience was becoming more and more hopeless.

III
Disappointments

M Y THIRD encounter with Stalin came in early 1948. This was the most significant encounter, for it took place on the eve of the rift between the Soviet and the Yugoslav leaders. It was preceded by significant events and changes in Yugoslav-Soviet relations.

Relations between the Soviet Union and the West had already assumed the contours of the Cold War between two blocs. The key events leading to this, in my opinion, were the Soviet rejection of the Marshall Plan, the civil war in Greece, and the creation by some Communist parties of an Information Bureau, the Cominform. Yugoslavia and the Soviet Union were the only two East European countries that were decisively against the Marshall Plan—the former largely out of revolutionary dogmatism, and the latter for fear that American economic aid might shake up the empire it had so recently acquired militarily.

As Yugoslav delegate to the Congress of the Communist Party of France in Strasbourg, I found myself in Paris just at the time Molotov was having conversations with the representatives of the Western states regarding the Marshall Plan. Molotov received me in the Soviet Embassy, and we agreed on boycotting the Marshall Plan, and also in our criticism of the French Party, with its so-called "national line." Molotov was especially interested in my impressions of the Congress, and he remarked about the periodical *La Nouvelle Democratie,* of which

Duclos was the editor and which purported to express the united views of the Communist parties: "That isn't what was needed and what ought to be done."

Regarding the Marshall Plan, Molotov wondered whether a conference should not be called in which the Eastern countries would also participate, but only for propaganda reasons, with the aim of exploiting the publicity and then walking out on the conference at a convenient moment. I was not enthusiastic about this variation either, though I would not have opposed it had the Russians insisted; such was the stand taken by my country's Government. However, Molotov received a message from the Politburo in Moscow that he should not agree even to this.

Immediately upon my return to Belgrade I learned that a conference of East European countries was to be held in Moscow to take a stand with respect to the Marshall Plan. I was designated to represent Yugoslavia. The real aim of the conference was to bring collective pressure to bear on Czechoslovakia, whose Government was not against participating in the Marshall Plan. The Soviet plane was already waiting at the Belgrade airfield, but I did not fly the next day, for a telegram arrived from Moscow stating that there was no need for the conference —the Czechoslovak Government had abandoned its original stand.

That same conformity with the Soviet Union, though for reasons other than the Soviet Union's, manifested itself also in the creation of the Cominform. The idea that it was necessary to create some agency that could facilitate the co-ordination and exchange of views among the Communist parties had been discussed as early as 1946; Stalin, Tito, and Dimitrov had talked about it in the spring of the same year. However, its realization had been postponed for many reasons, mostly, to be sure, because everything depended on the Soviet leaders' judgment of when the time was ripe. It ripened in the fall of 1947, most probably in connection with the Soviet rejection of the Marshall Plan and the solidification of Soviet domination over Eastern Europe.

At the founding meeting—in western Poland, that is, on former German territory—the only two delegations that were decidedly for the Cominform were the Yugoslav and the Soviet. Gomulka was opposed, cautiously but unequivocally holding out for the "Polish path to socialism."

In connection with this, I might mention as a curiosity that it was Stalin who thought up the name of the Cominform's organ, *For a Lasting Peace—For a People's Democracy,* with the idea that the Western press would have to repeat the slogan each time it quoted something from it. But Stalin's expectation was not fulfilled: because of the length and transparent propaganda nature of its name, the newspaper was—as though for spite—most frequently referred to simply as "the organ of the Cominform." Stalin also decided in the end where the seat of the Cominform was to be. The delegates had agreed on Prague. The Czech representative, Slansky, hurried to Prague by car that evening to consult Gottwald about this. But that night Zhdanov and Malenkov talked with Stalin (for not even in that remote pension and distant location did they fail to have a direct telephone connection with Moscow), and though Gottwald was reluctant to agree, Stalin ordained that the seat should be in Belgrade.

This double-dealing was also going on in the depths of Yugoslav-Soviet relations: on the surface, complete political and, especially, ideological agreement, but in reality divergent practices and judgments.

When a rather broad delegation of the top Yugoslav leaders—Tito, Ranković, Kidrič, Nešković—sojourned in Moscow in the spring of 1946, relations between the two leading groups assumed a more than cordial appearance. Stalin embraced Tito, referred to his European-wide role, and flagrantly belittled the Bulgars and Dimitrov. But soon after there came the tension and discord over joint-stock companies.

The subterranean friction went on constantly. Invisible to the non-Communist world, it broke out in closed Party councils, over recruiting for the Soviet Intelligence Service; which was particularly inconsiderate with respect to the state and Party apparatus. It broke out also in the sphere of ideology especially because of Soviet disparagement of the Yugoslav revolution. The Soviet representatives swallowed with obvious distaste the Yugoslavs' ranking of Tito next to Stalin, and they were particularly sensitive about Yugoslavia's independent association with the other East European countries and the growth of her prestige among them.

The friction soon carried over into economic relations, especially when it became obvious to the Yugoslavs that, apart from their ordinary commercial ties, they could not count on Soviet aid in carrying out their five-year plan. Detecting resistance, Stalin stressed that the use of joint-stock companies was not good among friendly and allied countries and promised to furnish all possible aid, but at the same time his traders exploited the economic advantage they gained as a result of exacerbated Yugoslav-Western relations and as a result of the illusory Yugoslav view of the USSR as an unselfish and unhegemonistic state.

Except for Albania, Yugoslavia had been the only East European country to free itself from the Nazi invasion and at the same time carry out a domestic revolution without the decisive help of the Red Army. It had gone the farthest in effecting a social transformation, and yet it was also situated in what was in days to come the most exposed salient in the Soviet bloc. In Greece a civil war was being fought. Yugoslavia had been charged in the United Nations with giving it material aid and inspiring it; while Yugoslav relations with the West, and especially with the United States, were strained to the breaking point.

When I think back, it seems to me that the Soviet Government not only looked with satisfaction at this sharpening of Yugoslav-Western relations but even incited it, taking care, of course, not to go beyond the limits of its own interests and possibilities. Molotov almost embraced Kardelj in Paris after the shooting down of two American planes in Yugoslavia, though he also cautioned him against shooting down a third. The Soviet Government took no direct action with respect to the uprising in Greece, practically leaving Yugoslavia to face the music alone in the United Nations, nor did it undertake any-thing decisive to bring about an armistice—not until Stalin found it to his interest.

So, too, the designation of Belgrade as the seat of the Cominform was, on the surface, recognition for the Yugo-slav revolution. Behind it lay the secret Soviet intention to lull to sleep the Yugoslav leaders with revolutionary self-satisfaction and to subordinate Yugoslavia to some supposed international Communist solidarity—in fact, to the hegemony of the Soviet state, or, rather, to the in-satiable demands of the Soviet political bureaucracy.

It is time something was said about Stalin's attitude toward revolutions, and thus toward the Yugoslav revo-lution. Because Moscow abstained, always in decisive moments, from supporting the Chinese, Spanish, and in many ways even the Yugoslav revolutions, the view prevailed, not without reason, that Stalin was generally against revolutions. This is, however, not entirely cor-rect. He was opposed only conditionally, that is, to the degree to which the revolution went beyond interests of the Soviet state. He felt instinctively that the creation of revolutionary centers outside of Moscow could endanger its supremacy in world Communism, and of course that is what actually happened. That is why he helped revolu-tions only up to a certain point—up to where he could control them—but he was always ready to leave them in the lurch whenever they slipped out of his grasp. I main-tain that not even today is there any essential change in this respect in the policy of the Soviet Government.

A man who had subjected all activities in his own country to his views and to his personality, Stalin could not behave differently outside. Having identified domestic progress and freedom with the interests and privileges of a political party, he could not act in foreign affairs other than as a hegemonist. As with everyone, handsome is as handsome does. He became himself the slave of the despotism, the bureaucracy, the narrowness, and the servility that he imposed on his country.

It is indeed true that no one can take freedom from another without losing his own.

2

The occasion for my departure to Moscow was the divergence between the policy of Yugoslavia and that of the USSR toward Albania. In late December of 1947 there came from Moscow a dispatch in which Stalin demanded that someone of the Yugoslav Central Committee—he spoke of me only by name—come in order to bring into harmony the policies of the two Governments vis-à-vis Albania.

The disharmony made itself felt in various ways, most visibly after the suicide of Naku Spiru, a member of the Albanian Central Committee.

A linkage between Yugoslavia and Albania had been developing in all fields. Yugoslavia was sending to Albania experts of all kinds in ever increasing numbers. Food was shipped to Albania, though Yugoslavia itself suffered a shortage. The creation of joint-stock companies had begun. Both Governments agreed in principle that

Albania ought to unite with Yugoslavia, which would have solved the question of the Albanian minority in Yugoslavia.

The conditions that the Yugoslav Government presented the Albanian were far more favorable and just for the Albanians than those, by comparison, that the Soviet Government offered to the Yugoslavs. Apparently, however, the problem lay not in the degree of justice but in the very nature of these relations. A part of the Albanian leadership was intimately and secretly against the Yugoslav approach.

Naku Spiru—slight, frail, very sensitive, with a fine intellect—directed the economic affairs of the Albanian Government at the time and was the first to rebel against Yugoslavia, demanding that Albania develop independently. His stand provoked a sharp reaction not only in Yugoslavia but in the Albanian Central Committee as well. He was especially opposed by Koči Xoxe, Albanian Minister of the Interior, who was later shot on the charge that he was pro-Yugoslav. A worker from southern Albania and a veteran revolutionary, Xoxe enjoyed the reputation of being the most stable Party man despite the fact that Enver Hoxha—an undoubtedly better-educated and far more agile personality—was Secretary General of the Party and Premier of the Government. Hoxha, too, joined in the criticism against Spiru, even though his actual position remained unclear. Poor Spiru, finding himself isolated and charged with chauvinism and probably on the brink of being expelled from the Party, killed himself. With his death he started something he never

would have imagined—the worsening of Yugoslav-Albanian relations.

To be sure, the affair was hushed up before the public.
Later, after the open break with Yugoslavia in 1948,
Enver Hoxha placed Spiru on a pedestal as a national
hero. But in the summits of both countries the affair left
a bad impression which could not be dispelled by asser-
tions concerning Spiru's cowardice, petty bourgeois spirit,
and the like, which always abound in the Communist
arsenal of clichés.

The Soviet Government was excellently informed both
about the real causes of Spiru's death and about all of
Yugoslavia's activities in Albania. Her Mission in Tirana
grew more and more numerous. Besides, relations among
the three Governments—the Soviet, Albanian, and Yugoslav—were such that the last two did not particularly
conceal their relations from the first, though it should
also be said that the Yugoslav Government did not consult the Soviet with respect to the details of its policy.

Soviet representatives made ever more frequent complaints about certain Yugoslav measures in Albania,
while an ever greater closeness was observed between
the group around Hoxha and the Soviet Mission. Every
once in a while a complaint by this or that Soviet representative came to the surface: Why were the Yugoslavs
forming joint-stock companies with the Albanians when
they refused to form the same in their own country with
the USSR? Why were they sending their instructors to the
Albanian Army when they had Soviet instructors in their
own? How could Yugoslavs provide experts for the de-

velopment of Albania when they themselves were seeking
experts from abroad? How was it that all of a sudden
Yugoslavia, itself poor and underdeveloped, intended
to develop Albania?

Along with these divergencies between the Soviet and
the Yugoslav Governments, Moscow's tendency to re-
place Yugoslavia's position in Albania became all the
more evident, which seemed extremely unjust to the
Yugoslavs in view of the fact that it was not the USSR
that proposed to unite with Albania, nor was the USSR
even a bordering neighbor of Albania's. The turning of
the Albanian leaders to the Soviet Union became increasingly evident and found ever more lively expression
in their propaganda.

The Soviet Government's invitation to remove disagreement over Albania was accepted with both hands
in Belgrade, though it has remained unclear to this day
why Stalin emphasized that he wanted precisely me to
come to Moscow.

It seems to me that he was led by two reasons. I probably must have given him the impression of being a forth-
right and candid man. I expect that I was considered
such among the Yugoslav Communists too. As such I
was suitable for a straightforward discussion over a complicated and very sensitive question. However, I also
believe that he had the intention of winning me over
in order to split and to subordinate the Yugoslav Central Committee. He already had Hebrang and Žujović
on his side. But Hebrang had been thrown out of the
Central Committee and placed under secret investiga-
tion because of his unexplained behavior while in prison

during the war. Žujović was a prominent figure, but even as a member of the Central Committee he did not belong to the inner circle that had formed around Tito in the course of the struggle for the unity of the Party and during the revolution itself.

During Tito's stay in Moscow in 1946, when he told Stalin that I suffered from headaches, Stalin had invited me to visit him in the Crimea for a rest cure. But I did not go, largely because Stalin's invitation had not been made again through the Embassy, and so I took it to be a polite gesture, made simply because the conversation had turned to me.

Thus I set out for Moscow—on January 8, if I remember correctly, and certainly not far from that date—with ambiguous feelings: I was flattered that Stalin had invited me specifically, but I also had vague, unutterable suspicions that this was neither by chance nor with pure intentions with regard to Tito and the Yugoslav Central Committee.

I received no special orders or instructions in Belgrade, nor were any instructions necessary, inasmuch as I was a member of the inner circle of leaders and *au courant* on Albanian-Yugoslav relations. The stand had already been formulated that Soviet representatives should not hinder the already announced policy of Yugoslav-Albanian unification by their tactless actions or by taking a different line.

Representatives of the Yugoslav Army took advantage of a good opportunity to send with me their own delegation, which was to present requests for munitions and for the development of our war industry. This delegation included the then Chief of the General Staff, Koča Popović, and the head of the Yugoslav war industry, Mijalko Todorović. Svetozar Vukmanović-Tempo, then director of political administration in the army, also traveled with us, in order to acquaint himself with the experience of the Red Army in that area.

We set out by train for Moscow, in good spirits and in even better faith. And also with the set view that Yugoslavia should solve its problems in its own way and largely through its own resources.

3

This view was aired even before it should have been, at a dinner in the Yugoslav Embassy in Bucharest which was attended by Anna Pauker, the Rumanian Foreign Minister, and several Rumanian officials.

All the Yugoslavs, except Ambassador Golubović, who later emigrated as an adherent of Moscow, more or less openly brought out that the Soviet Union could not be an absolute model in "the building of socialism," for the situation had changed and conditions and circumstances differed in the individual countries of Eastern Europe. I noticed that Anna Pauker was carefully silent, or else agreed with something reluctantly, and tried to avoid talking about such sensitive questions. One of the Rumanians—I believe it was Bodnaraš—opposed our views, and another—his name I have unfortunately forgotten—cordially agreed with us. I regarded a conversation of this sort awkward, for I was convinced that every word would reach the ears of the Russians and they would be unable to understand them as being anything but "anti-Soviet" —synonym of all the evils of this earth. At the same time, however, I could not retract my stand. Thus I sought to tone down these views, stressing the merits of the USSR and the theoretical significance of the Soviet experience. But all this was of hardly any use, for I myself had stressed that everyone ought to blaze his own path according to his own concrete circumstances. Nor could the awkwardness be dispelled. I had a premonition; I knew that the Soviet leaders had no feeling for nuances and compromises, especially not within their own Communist ranks.

Though we were only passing through Rumania, we found reason for our criticism everywhere. First, as to the relations between the Soviet Union and the other East European countries: these countries were still being held under actual occupation, and their wealth was being extracted in various ways, most frequently through joint-stock companies in which the Russians barely invested anything except German capital, which they had simply declared a prize of war. Trade with these countries was not conducted as elsewhere in the world, but on the basis of special arrangements according to which the Soviet Government bought at lower and sold at higher than world prices. Only Yugoslavia was an exception. We knew all that. And the spectacle of misery as well as the consciousness of impotence and subservience among the Rumanian authorities could only heighten our indignation.

We were most taken aback by the arrogant attitude of

the Soviet representatives. I remember how horrified
we were at the words of the Soviet Commander in Iaşi:
"Oh, this dirty Rumanian Iaşi! And these Rumanian
corn-pone eaters (*mamalizhniki*)!" He also repeated
Ehrenburg's and Vishinsky's bon mot, which was aimed
at the corruption and stealing in Rumania: "They are
not a nation, but a profession!"

Especially in that mild winter, Iaşi was truly a sprawl-
ing backwater of a Balkan town whose beauties—its hills,
gardens, and terraces—could be detected only by the experienced eye. Yet we knew that Soviet towns looked
hardly better, if not indeed worse. It was this attitude
of a "superior race" and the conceit of a great power
that angered us the most. The obliging and deeply
respectful Russian attitude toward us not only accentu-
ated the abasement of the Rumanians all the more, but
it inflated our pride in our own independence and in our
freedom to reason.

We had already accepted it as a fact of life that such
relations and attitudes as existed toward the Rumanians
were "possible even in socialism" because "Russians are
like that"—backward, long isolated from the rest of the
world, and dead to their revolutionary traditions.

We bored ourselves in Iaşi a few hours, until the Soviet
train with the Soviet Government's car arrived for us,
accompanied, to be sure, by the inevitable Captain
Kozovsky, for whom the Yugoslavs continued to be his
specialty in the Soviet State Security. This time he was
less unreserved and sunny than before, probably only
because he was now faced by ministers and generals. An
intangible, undefinable, cold officialism intruded itself

in the relations between ourselves and our Soviet "comrades."

Our sarcastic comments did not spare even the railroad car in which we traveled, and which deserved no better despite its comfortable accommodations, the excellent food, and the good service. We regarded as comical the huge brass handles, the old-fashioned fussiness of the décor, and a toilet so lofty that one's legs dangled in mid-air. Was all this necessary? Does a great state and a sovereign power have to show off? And what was most grotesque of all in that car, with its pomp of tsarist days, was the fact that the conductor kept, in a cage in his compartment, a chicken which laid eggs. Poorly paid and miserably clothed, he apologized: "What is one to do, Comrades? A workingman must make out as best he can. I have a big family—and life is hard."

Though the Yugoslav railroad system could hardly boast of accuracy either, here no one got excited over a tardiness of several hours. "We'll get there," one of the conductors would simply reply. Russia seemed to confirm the unchangeability of its human and national soul; all its essential qualities militated against the pace of industrialization and the omnipotence of management.

The Ukraine and Russia, buried in snow up to the eaves, still bore the marks of the devastation and horrors of war—burned-down stations, barracks, and the sight of women, on the subsistence of hot water (*kipiatok*) and a piece of rye bread, wrapped in shawls, clearing tracks.

This time, too, only Kiev left an impression of discreet beauty and cleanliness, culture and a feeling for style and taste, despite its poverty and isolation. Because it was night, there was no view of the Dnieper and the plains merging with the sky. Still it all reminded one of Belgrade—the future Belgrade, with a million people and built with diligence and harmony. We stopped in Kiev only briefly, to be switched to the train for Moscow. Not one Ukrainian official met us. Soon we were on our way into a night white with snow and dark with sorrow. Only our car sparkled with the brilliance of comfort and abundance in this limitless desolation and poverty.

4

Just a few hours after our arrival in Moscow we were deep in a cordial conversation with the Yugoslav Ambassador, Vladimir Popović, when the telephone on his desk rang. The Soviet Ministry of Foreign Affairs was asking if I was tired, for Stalin wished to see me immediately, that same evening. Such haste is unusual in Moscow, where foreign Communists have always waited long, so that a saying circulated among them: It is easy to get to Moscow but hard to get out again. To be sure, even if I had been tired, I would have accepted Stalin's invitation most willingly. Everyone in the delegation regarded me with enthusiasm, though also not without envy, and Koča Popović and Todorović kept reminding me not to forget why they, too, had come along, even though I had taken advantage of our traveling together to acquaint myself in detail with their requests.

My joy over the impending meeting with Stalin was sober and not quite pure precisely because of the haste with which it had come. This misgiving never left me the whole night that I spent with him and other Soviet leaders.

As usual, at about nine o'clock in the evening they took me to the Kremlin, to Stalin's office. Gathered there were Stalin, Molotov, and Zhdanov. The last, as was known to me, had charge in the Politburo of maintaining relations with foreign parties.

After the customary greetings, Stalin immediately got down to business: "So, members of the Central Committee in Albania are killing themselves over you! This is very inconvenient, very inconvenient."

I began to explain: Naku Spiru was against linking Albania with Yugoslavia; he isolated himself in his own Central Committee. I had not even finished when, to my surprise, Stalin said: "We have no special interest in Albania. We agree to Yugoslavia swallowing Albania! . . ." At this he gathered together the fingers of his right hand and, bringing them to his mouth, he made a motion as if to swallow them.

I was astonished, almost struck dumb by Stalin's manner of expressing himself and by the gesture of swallowing, but I do not know whether this was visible on my face, for I tried to make a joke of it and to regard this as Stalin's customary drastic and picturesque manner of expression. Again I explained: "It is not a matter of swallowing, but unification!"

At this Molotov interjected: "But that is swallowing!"

And Stalin added, again with that gesture of his: "Yes, yes. Swallowing! But we agree with you: you ought to swallow Albania—the sooner the better."

Despite this manner of expression, the whole atmosphere was cordial and more than friendly. Even Molotov expressed that bit about swallowing with an almost humorous amiability which was hardly usual with him.

I approached a *rapprochement* and unification with Albania with sincere and, of course, revolutionary motives. I considered, as did many others, that unification— with the truly voluntary agreement of the Albanian leaders—would not only be of direct value to both Yugoslavia and Albania, but would also finally put an end to the traditional intolerance and conflict between Serbs and Albanians. Its particular importance, in my opinion, lay in the fact that it would make possible the amalgamation of our considerable and compact Albanian minority with Albania as a separate republic in the Yugoslav-Albanian Federation. Any other solution to the problem of the Albanian national minority seemed impracticable to me, since the simple transfer of Yugoslav territories inhabited by Albanians would give rise to uncontrollable resistance in the Yugoslav Communist Party itself.

I had for Albania and the Albanians a special predilection which could only strengthen the idealism of my motivations: The Albanians, especially the northern ones, are by mentality and way of life akin to the Montenegrins from whom I spring, and their vitality and determination to maintain their independence has no equal in human history.

Though it did not even occur to me to differ with the view of my country's leaders and to agree with Stalin, still Stalin's interjections for the first time confronted me with two thoughts. The first was the suspicion that something was not right about Yugoslavia's policy toward Albania, and the other was that the Soviet Union had united with the Baltic countries by swallowing them. It was Molotov's remark that directly reminded me of this.

Both thoughts merged into one—into a feeling of discomfort.

The thought that there might be something obscure and inconsistent about Yugoslav policy toward Albania did not, however, cause me to admit that this policy was one of "swallowing." Yet it did strike me that this policy did not correspond with the will and the desires of the Albanian Communists, which, for me, as a Communist, were identical with the aspirations of the Albanian people. Why *did* Spiru kill himself? He was not "petty bourgeois" and "burdened by nationalism" as much as he was a Communist and a Marxist. And what if the Albanians wished, as we did vis-à-vis the Soviet Union, to have their own separate state? If the unification were carried out despite Albanian wishes and by taking advantage of their isolation and misery, would this not lead to irreconcilable conflicts and difficulties? Ethnically peculiar and with an ancient identity, the Albanians as a nation were young and hence filled with an irrepressible and still unfulfilled national consciousness. Would they not consider unification as the loss of their independence, as a rejection of their individuality?

As for the other thought—that the USSR had swallowed the Baltic states—I linked it with the first, repeating, convincing myself: We Yugoslavs do not wish, do not dare, to take that road to unification with Albania, nor is there any immediate danger that some imperialistic power, such as Germany, might bring pressure on Albania and use it as a base against Yugoslavia.

But Stalin brought me back to reality. "And what about Hoxha, what is he like in your opinion?"

I avoided a direct and clear answer, but Stalin expressed precisely the same opinion of Hoxha as the Yugoslav leaders had acquired. "He is a petty bourgeois, inclined toward nationalism? Yes, we think so too. Does it seem that the strongest man there is Xoxe?"

I confirmed his leading questions.

Stalin ended the conversation about Albania, which lasted barely ten minutes: "There are no differences between us. You personally write Tito a dispatch about this in the name of the Soviet Government and submit it to me by tomorrow."

Afraid that I had not understood, I sounded him out, and he repeated that I was to write the dispatch to the Yugoslav Government in the name of the Soviet Government.

At that moment I took this to be a sign of special confidence in me and as the highest expression of agreement with the Yugoslav policy toward Albania. However, while writing the dispatch the next day, the thought occurred to me that it might someday be used against my country's Government, and so I formulated it carefully and very briefly, something like this: Djilas arrived in Moscow yesterday and, at a meeting held with him on the same day, there was expressed complete agreement between the Soviet Government and Yugoslavia concerning the question of Albania. That dispatch was never sent to the Yugoslav Government, nor was it ever used against it in later clashes between Moscow and Belgrade.

The rest of the conversation did not last long either and revolved idly around such uneventful questions as the location of the Cominform in Belgrade and its newspaper, Tito's health, and the like.

However, I seized an opportune moment and raised the question of supplies for the Yugoslav Army and our war industry. I stressed that we frequently encountered difficulties with Soviet representatives because they refused to give us this or that, using "military secrets" as an excuse. Stalin rose shouting, "We have no military secrets from you. You are a friendly socialist country—we have no military secrets from you."

He then went to his desk, called Bulganin on the phone, and gave a short order: "The Yugoslavs are here, the Yugoslav delegation—they should be heard immediately."

The whole conversation in the Kremlin lasted about a half hour, and then we set out for Stalin's villa for dinner.

5

We seated ourselves in Stalin's automobile, which seemed to me to be the same as the one in which I rode with Molotov in 1945. Zhdanov sat in back to my right, while Stalin and Molotov sat in front of us on the folding seats. During the trip Stalin turned on a little light on the panel in front of him under which hung a pocket watch—it was almost ten o'clock—and I observed directly in front of me his already hunched back and the bony gray nape of his neck with its wrinkled skin above the stiff marshal's collar. I reflected: Here is one of the most powerful men of today, and here are his associates; what a sensational catastrophe it would be if a bomb now exploded in our midst and blew us all to pieces! But this thought was only fleeting and ugly and so unexpected even to myself that it horrified me. With a sad affection, I saw in Stalin a little old grandfather who, all his life, and still now, looked after the success and happiness of the whole Communist race.

While waiting for the others to gather together, Stalin, Zhdanov, and I found ourselves in the entrance hall of the villa, by the map of the world. I again glanced at the blue pencil mark that encircled Stalingrad—and again Stalin noticed it; I could not fail to observe that my scrutiny pleased him. Zhdanov also noticed this exchange of glances, joined us, and remarked, "The beginning of the Battle of Stalingrad."

But Stalin said nothing to that.

If I remember well, Stalin began to look for Königsberg, for it was to be renamed Kaliningrad—and in so doing we came across places around Leningrad that still bore German names from the time of Catherine. This caught Stalin's eye and he turned to Zhdanov, saying curtly: "Change these names—it is senseless that these places still bear German names!" At this Zhdanov pulled out a small notebook and recorded Stalin's order with a little pencil.

After this Molotov and I went to the toilet, which was located in the basement of the villa. It contained several stalls and urinals. Molotov began to unbutton his pants even as we walked, commenting: "We call this unloading before loading!" Thereupon I, a long-time resident of prisons, where a man is forced to forget about modesty, felt ashamed in the presence of Molotov, an older man, entered a stall and shut the door.

After this both of us proceeded to the dining room, where Stalin, Malenkov, Beria, Zhdanov, and Voznesensky were already gathered. The last two are new personae in these memoirs.

Zhdanov, too, was rather short, with a brownish clipped mustache, high forehead, pointed nose, and a sickly red face. He was educated and was regarded in the Politburo as a great intellectual. Despite his well-known narrowness and dogmatism, I would say that his knowledge was not inconsiderable. Although he had some knowledge of everything, even music, I would not say that there was a single field that he knew thoroughly—a typical intellectual who became acquainted with and picked up knowledge of other fields through Marxist literature. He was also a cynic, in an intellectual way, but all the uglier for this because behind the intellectualism one unmistakably sensed the potentate who was "magnanimous" toward men of the spirit and the pen. This was the period of the "Decrees"—decisions by the Soviet Central Committee concerning literature and other branches of the arts which amounted to a violent attack against even those minimal freedoms in the choice of subject and form that had survived (or else had been snatched from) bureaucratic Party control during the war. I remember that that evening Zhdanov recounted as the latest joke how his criticism of the satirist Zoshchenko had been taken in Leningrad: They simply took away Zoshchenko's ration coupons and did not give them back to him until after Moscow's magnanimous intervention.

Voznesensky, the Chairman of the Planning Commission of the USSR, was barely past forty—a typical Russian, blond and with prominent cheekbones, a rather high forehead, and curly hair. He gave the impression of being an orderly, cultured, and above all withdrawn man, who said little and always had a happy inward smile. I had previously read his book on the Soviet economy during the war, and it gave me the impression that the author was a conscientious and thoughtful man. Later that book was criticized in the USSR, and Voznesensky was liquidated for reasons that have remained undisclosed to this day.

I was well acquainted with Voznesensky's older brother, a university professor who had just been named Minister of Education in the Russian Federation. I had had some very interesting discussions with the elder Voznesensky at the time of the Panslavic Congress in Belgrade, in the winter of 1946. We had agreed not only with respect to the narrowness and bias of the prevailing theories of "socialist realism," but also concerning the appearance of new phenomena in socialism (that is, communism) with the creation of the new socialist countries and with changes in capitalism which had not yet been discussed theoretically. It is probable that his handsome contemplative head also fell in the senseless purges.

The dinner began with someone—it seems to me that it was Stalin himself—proposing that everyone guess how many degrees below zero it was, and that everyone be punished by being made to drink as many glasses of vodka as the number of degrees he guessed wrong. Luckily, while still at the hotel, I had looked at the thermometer, and I added to the number to allow for the temperature drop during the night, so that I missed by only one degree. I remember that Beria missed by three, remarking that he had done so on purpose so that he might drink more glasses of vodka.

Such a beginning to a dinner forced upon me a heretical thought: These men shut up in a narrow circle were capable of inventing even more senseless reasons for drinking vodka—the length of the dining room in feet or of the table in inches. And who knows, maybe that's what they do! At any rate, this apportioning of the number of vodka glasses according to the temperature reading suddenly brought to my mind the confinement, the inanity and senselessness of the life these Soviet leaders were living gathered about their superannuated chief even as they played a role that was decisive for the human race. I recalled that the Russian tsar Peter the Great likewise held such suppers with his assistants at which they gorged and drank themselves into a stupor while ordaining the fate of Russia and the Russian people.

This impression of the vacuity of such a life did not recede but kept recurring during the course of the dinner despite my attempts to suppress it. It was especially strengthened by Stalin's age, by conspicuous signs of his senility. No amount of respect and love for his person, which I stubbornly nurtured inside myself, was able to erase that realization from my consciousness.

There was something both tragic and ugly in his senility. The tragic was invisible—these were the reflections in my head regarding the inevitability of decline in even so great a personality. The ugly kept cropping up all the time. Though he had always enjoyed eating well, Stalin now exhibited gluttony, as though he feared that there would not be enough of the desired food left for him. On the other hand, he drank less and more cautiously, as though measuring every drop—to avoid any ill effects.

His intellect was in even more apparent decline. He liked to recall incidents from his youth—his exile in Siberia, his childhood in the Caucasus; and he would compare everything recent with something that had already happened: "Yes, I remember, the same thing. . . ."

It was incomprehensible how much he had changed in two or three years. When I had last seen him, in 1945, he was still lively, quick-witted, and had a pointed sense of humor. But that was during the war, and it had been, it would seem, Stalin's last effort and limit. Now he laughed at inanities and shallow jokes. On one occasion he not only failed to get the political point of an anecdote I told him in which he outsmarted Churchill and Roosevelt, but I had the impression that he was offended, in the manner of old men. I perceived an awkward astonishment on the faces of the rest of the party.

In one thing, though, he was still the Stalin of old: stubborn, sharp, suspicious whenever anyone disagreed with him. He even cut Molotov, and one could feel the tension between them. Everyone paid court to him, avoiding any expression of opinion before he expressed his, and then hastening to agree with him.

As usual, they hopped from subject to subject—and I shall proceed likewise in my account.

Stalin spoke up about the atom bomb: "That is a powerful thing, pow-er-ful!" His expression was full of admiration, so that one was given to understand that he would not rest until he, too, had the "powerful thing." But he did not mention that he had it already or that the USSR was working on it.

On the other hand, when Kardelj and I met with Dimitrov in Moscow a month later, Dimitrov told us as if in confidence that the Russians already had the atom bomb, and an even better one than the Americans', that is, the one exploded over Hiroshima. I maintain that this was not true, but that the Russians were just on the way to making an atom bomb. But these are the facts, and I cite them.

Both that night and again soon after, in a meeting with the Bulgarian and Yugoslav delegations, Stalin stressed that Germany would remain divided: "The West will make Western Germany their own, and we shall turn Eastern Germany into our own state."

This thought of his was new, but understandable; it proceeded from the whole trend of Soviet policy in Eastern Europe and toward the West. I could never understand the statements by Stalin and the Soviet leaders, made before the Bulgars and the Yugoslavs in the spring of 1946, that all of Germany must be ours, that is, Soviet, Communist. I asked one of those present how the

Russians meant to bring this about. He replied, "I don't know myself!" I suspect that not even those who made the statements actually knew how but were caught up by the flush of military victories and by their hopes for the economic and other dissolution of Western Europe.

Toward the end of the dinner Stalin unexpectedly asked me why there were not many Jews in the Yugoslav Party and why these few played no important role in it. I tried to explain to him that there were not many Jews in Yugoslavia to begin with, and most belonged to the middle class. I added, "The only prominent Communist Jew is Pijade, and he regards himself as being more of a Serb than a Jew."

Stalin began to recall: "Pijade, short, with glasses? Yes, I remember, he visited me. And what is his position?"

"He is a member of the Central Committee, a veteran Communist, the translator of *Das Kapital*," I explained.

"In our Central Committee there are no Jews!" he broke in, and began to laugh tauntingly. "You are an anti-Semite, you, too, Djilas, you, too, are an anti-Semite!"

I took his words and laughter to mean the opposite, as I should have—as the expression of his own anti-Semitism and as a provocation to get me to declare my stand concerning the Jews, particularly Jews in the Communist movement. I laughed softly and kept still, which was not difficult for me inasmuch as I have never been an anti-Semite and I divided Communists solely into the good and the bad. Stalin himself quickly abandoned this slippery subject, being content with his cynical provocation.

At my left sat the taciturn Molotov, and at my right the loquacious Zhdanov. The latter told of his contacts with the Finns and admiringly emphasized their exactitude in delivering reparations: "Everything on time, expertly packed, and of excellent quality."

He concluded, "We made a mistake in not occupying Finland. Everything would have been set up if we had."

Molotov: "Akh, Finland—that is a peanut."

At that very time Zhdanov was holding meetings with composers and preparing a "decree" on music. He liked operas and asked me in passing, "Do you have opera in Yugoslavia?"

Surprised at his question, I replied, "In Yugoslavia operas are being presented in nine theaters!" At the same time I thought: How little they know about Yugoslavia. Indeed, it is not noticeable that it even interests them except as a given geographic location.

Zhdanov was the only one who was drinking orangeade. He explained to me that he did this because of his bad heart. I asked him, "How serious is your illness?"

With a restrained smile he replied with his customary mockery, "I might die at any moment, and I might live a very long time." He certainly evinced an exaggerated sensitivity, and he reacted quickly and too easily.

A new five-year plan had just been promulgated. Without turning to anyone in particular Stalin announced that the teachers' salaries ought to be increased. And then to me: "Our teachers are very good, but their salaries are low—we must do something."

Everyone uttered a few words of agreement while I recalled, not without bitterness, the low salaries and wretched conditions of Yugoslav cultural workers and my impotence to help them.

Voznesensky kept silent the whole time; he comported himself like a junior among seniors. Stalin addressed him directly only with this one question: "Could means be obtained outside of the Plan for the construction of the Volga-Don Canal? A very important job! We must find the means! A terribly important job from the military point of view as well: in case of war they might drive us out of the Black Sea—our fleet is weak and will go on being weak for a long time. What would we do with our ships in that case? Imagine how valuable the Black Sea Fleet would have been during the Battle of Stalingrad if we had had it on the Volga! That canal is of first-class, first-class importance."

Voznesensky agreed that the means could be found, took out a little notebook and made a note of it.

I had long been interested in two questions—almost privately—and I wished to ask Stalin for his opinion. One was in the field of theory: neither in Marxist literature nor anywhere else could I ever find an explanation of the difference between "people" and "nation." Since Stalin had long been reputed among Communists to be an expert on the nationalities question, I sought his opinion, pointing out that he had not treated this in his book on the nationalities question, which had been published even before the First World War and since then was considered the authoritative Bolshevik view.

At my question Molotov first joined in: " 'People' and 'nation' are both the same thing."

But Stalin did not agree. "No, nonsense! They are different!" And he began to explain simply: " 'Nation'? You already know what it is: the product of capitalism with given characteristics. And 'people'—these are the working-men of a given nation, that is, workingmen of the same language, culture, customs."

And concerning his book *Marxism and the National Question,* he observed: "That was Ilyich's—Lenin's view. Ilyich also edited the book."

The second question involved Dostoevsky. Since early youth I had considered Dostoevsky in many ways the greatest writer of the modern age, and I could never square within myself the Marxist attacks on him.

Stalin also answered this simply: "A great writer and a great reactionary. We are not publishing him because he is a bad influence on the youth. But, a great writer!"

We turned to Gorky. I pointed out that I regarded as his greatest work—both in method and in the depth of his depiction of the Russian Revolution—*The Life of Klim Samgin.* But Stalin disagreed, avoiding the subject of method. "No, his best things are those he wrote earlier: *The Town of Okurov,* his stories, and *Foma Gordeev.* And as far as the depiction of the Russian Revolution in *Klim Samgin* is concerned, there is very little revolution there and only a single Bolshevik—what was his name: Liutikov, Liutov?"

I corrected him: "Kutuzov—Liutov is an entirely different character."

Stalin concluded: "Yes, Kutuzov! The revolution is portrayed from one side, and inadequately at that; and from the literary point of view, too, his earlier works are better."

It was clear to me that Stalin and I did not understand one another and that we could not agree, though I had

had an opportunity to hear the opinions of significant littérateurs who, like himself, considered these particular works of Gorky his best.

Speaking of contemporary Soviet literature, I, as more or less all foreigners do, referred to Sholokhov's strength. Stalin observed: "Now there are better ones!"—and he cited two names, of which one belonged to a woman. Both were unknown to me.

I avoided a discussion of Fadeev's *Young Guard*, which even then was under attack for the insufficient "Partyness" of its heroes; also, Aleksandrov's *History of Philosophy*, which was criticized on quite opposite grounds— dogmatism, shallowness, banality.

It was Zhdanov who reported Stalin's observation on the book of love poems by K. Simonov: "They should have published only two copies—one for her, and one for him!" At which Stalin smiled demurely while the others roared.

The evening could not go by without vulgarity, to be sure, Beria's. They forced me to drink a small glass of *peretsovka*—strong vodka with pepper (in Russian, *perets* means pepper, hence the name for this drink). Sniggering, Beria explained that this liquor had a bad effect on the sex glands, and he used the most vulgar expressions in so doing. Stalin gazed intently at me as Beria spoke, ready to burst into laughter, but he remained serious on noticing how sour I was.

Even apart from this I could not dispel that conspicuous similarity between Beria and the Belgrade Royal Police official Vujković; it even grew to such proportions that I felt as though I was actually in the fleshy and damp clutches of Vujković-Beria.

However, I regarded as most important of all the atmosphere that permeated above and beyond the words during the course of the entire six hours of that dinner. Behind what was said, something more important was noticeable—something that ought to have been spoken, but that no one could or dared bring up. The forced conversation and the choice of topics made this something seem quite real, almost perceptible to the senses. I was even inwardly sure of its content: it was criticism of Tito and of the Yugoslav Central Committee. In that situation I would have regarded such criticism as tantamount to a recruiting of me on the part of the Soviet Government. Zhdanov was particularly energetic, not in any concrete, tangible way, but by injecting a certain cordiality, even intimacy into his conversation with me. Beria fixed me with his clouded green, gaping eyes while a self-conscious irony almost dripped down his square flabby mouth. Over them all stood Stalin—attentive, exceptionally moderate, and cold.

The mute gaps between topics became ever longer and the tension grew, both in and around me. I quickly worked out a strategy of resistance. Apparently it had been half consciously in the making inside of me even earlier. I would simply point out that I perceived no differences between the Yugoslav and Soviet leaders, that their aims were the same, and the like. A dumb, stubborn resistance welled inside of me, and though I had never before felt any inner vacillation, still I knew, knowing myself, that my defensive posture might easily turn into an offensive one if Stalin and the rest forced me into the moral dilemma of choosing between them and my conscience—or, under the circumstances, between their Party and mine, between Yugoslavia and the USSR. In order to prepare the ground, I referred to Tito and to my Central Committee several times in passing, but in a way that would not lead my interlocutors to launch into what they intended.

Stalin's attempt to introduce personal, intimate elements was in vain. Recalling his invitation in 1946, made via Tito, he asked me: "And why did you not come to the Crimea? Why did you refuse my invitation?"

I expected that question, and yet I was rather unpleasantly surprised that Stalin had not forgotten about it. I explained: "I waited for an invitation through the Soviet Embassy. I felt awkward about forcing myself and annoying you."

"Nonsense, no annoyance at all. You just didn't wish to come!" Stalin tested me.

But I drew back into myself—into chill reserve and silence.

And so nothing happened. Stalin and his group of cold, calculating conspirators—for I felt them to be so—certainly detected my resistance. This is just what I wanted. I had eluded them, and they did not dare provoke that resistance. They probably thought they had avoided a premature and thus erroneous step, but I became aware of that underhanded game and felt inside myself an inner, hitherto unknown, strength which was capable of rejecting even that by which I lived.

Stalin ended the dinner by raising a toast to Lenin's memory: "Let us drink to the memory of Vladimir Ilyich, our leader, our teacher—our all!"

We all stood and drank in mute solemnity, which, in our drunkenness we soon forgot, but Stalin continued to bear an earnest, grave, and even somber expression.

We left the table, but before we began to disperse, Stalin turned on a huge automatic record player. He even tried to dance, in the style of his homeland. One could see that he was not without a sense of rhythm. However, he soon stopped, with the resigned explanation, "Age has crept up on me and I am already an old man!"

But his associates—or, better said, courtiers—began to assure him, "No, no, nonsense. You look fine. You're holding up marvelously. Yes, indeed, for your age . . ."

Then Stalin turned on a record on which the coloratura warbling of a singer was accompanied by the yowling and barking of dogs. He laughed with an exaggerated, immoderate mirth, but on detecting incomprehension and displeasure on my face, he explained, almost as though to excuse himself, "Well, still it's clever, devilishly clever."

All the others remained behind, but were already preparing to leave. There was truly nothing more to say after such a long session, at which everything had been discussed except the reason why the dinner had been held.

6

We waited no more than a day or two before they invited us to the General Staff to present our requests. Earlier, while yet on board the train, I mentioned to Koča Popović and Mijalko Todorović that their requests seemed excessive and unrealistic to me. What I particularly could not get into my head was why the Russians would agree to build up the Yugoslav war industry when they did not wish to help seriously in developing our civilian industry, and it seemed even less likely to me that they would give us a war fleet when they lacked one themselves. The argument that it was all the same whether Yugoslavia or the USSR had a fleet on the Adriatic, since both were parts of a united Communist world, seemed all the more unconvincing to me precisely because of the cracks that were appearing in that very unity, not to speak of Soviet distrust of everything beyond their grasp and their unconcealed concern primarily for the interests of their own state. However, since all these requests had been elaborated and approved in Belgrade, there was nothing left for me but to stand by them.

The building of the General Staff was a pile whose external cheapness and artificiality they had in vain tried to compensate for internally by the lavish use of shrieking drapes and gilt. The meeting was presided over by Bulganin, surrounded by the highest military experts, among whom was also the Chief of the General Staff, Marshal Vasilevsky.

First I presented our needs generally, leaving the detailed presentation to Todorović and Popović. The Soviet officials did not commit themselves but they carefully went into our problems and took notes on everything. We left satisfied, convinced that matters had proceeded beyond a standstill and that the real concrete work would soon begin.

It indeed looked like it. Todorović and K. Popović were soon invited to further meetings. But everything came to an abrupt halt, and Soviet officials hinted that "complications" had set in and that we would have to wait.

It was clear to us that something was going on between Moscow and Belgrade, though we did not know exactly what, nor can I say that we were surprised. In any case, our critical attitude toward the Soviet reality and Moscow's stand toward Belgrade could only make the postponement of our talks all the more unbearable, especially since we found ourselves without anything to do, forced to kill time in conversation and by attending Moscow's old-fashioned but, as such, unsurpassed theaters.

None of the Soviet citizenry dared to visit us, for although we were from a Communist country, we still belonged to the category of foreigners, with whom citizens of the USSR could not associate, according to the letter of the law. All our contacts were limited to official channels in the Ministry of Foreign Affairs and in the Central Committee. That annoyed and offended us, all the more so since there were no such limitations in Yugoslavia, especially not for the representatives and citizens of the USSR. But this is what prompted us to draw critical conclusions.

Our criticism had not yet reached the point of generalization, but it abounded in examples taken from concrete reality. Vukmanović-Tempo had discovered faults in the army buildings which he did not conceal. In order to lessen our boredom, Koča Popović and I gave up our separate apartments in the Moskva Hotel, but we did not get a joint apartment until an "electrician" had put it in order, which we took to mean the installation of listening devices. Despite the fact that the Moskva was a new hotel and the largest, nothing in it worked as it should—it was cold, the faucets leaked, and the bathtubs, brought from Eastern Germany, could not be used because the drainage of water flooded the floor. The bathroom had no key, which gave Popović an occasion for his sparkling wit: The architect took into account that the key might get lost and he built the toilet near the door so that one could keep the door closed with one's foot.

I frequently recalled with envy my sojourn in the Metropole Hotel in 1944. Everything was old there, but in working order, and the superannuated help spoke English and French and demeaned themselves with grace and precision. But in the Moskva Hotel . . . One day I heard groaning in the bathroom. I came upon two workers there. One of them was repairing some fixtures on the ceiling, and the other was holding him up on his shoulders. "For heaven's sake, Comrades," said I, "why don't you get a ladder?" The workers complained, "We've asked the management for one lots of times, but no use —we always have a hard time like this."

Walking about we viewed "beautiful Moscow," most

of which was a big village, neglected and undeveloped. The chauffeur Panov, to whom I had sent a watch as a gift from Yugoslavia and with whom I had established a cordial relationship, found it impossible to believe that there were more cars in New York and Paris, although he did not hide his dissatisfaction with the quality of the new Soviet cars.

In the Kremlin, when we visited the imperial tombs, the girl guide spoke of "our tsars" with nationalist pathos. The superiority of the Russians was vaunted everywhere and assumed grotesque forms.

And so on down the line . . . At every step we discovered till then unnoticed aspects of the Soviet reality: backwardness, primitivism, chauvinism, a big-power complex, although accompanied by heroic and superhuman efforts to outgrow the past and to overtake the natural course of events.

Knowing that in the thick skulls of the Soviet leaders and political officials every least little criticism was transformed into an anti-Soviet attitude, we spontaneously entrenched ourselves in our own circle when in the presence of Russians. Since we were at the same time a political mission, we began to call each other's attention to anything "awkward" in our behavior or speech. This entrenchment began to assume an organized quality. I remember how, aware of the use of listening devices, we began to watch what we were saying in the hotel and in offices, and to turn on radios during conversations.

The Soviet representatives must have taken note of this. The tension and suspicion grew apace.

By that time Lenin's sarcophagus, which had been hidden somewhere in the interior during the war, had been brought back to Red Square. One morning we went to visit it. The visit itself would have had no importance had it, too, not provoked in me, as well as in the rest, a new and hitherto unknown resistance. As we descended slowly into the mausoleum, I saw how simple women in shawls were crossing themselves as though approaching the reliquary of a saint. I, too, was overcome by a feeling of mysticism, something forgotten from a distant youth. Moreover, everything was so arranged as to evoke just such a feeling in a man—the granite blocks, the stiff guards, the invisible source of light over Lenin, and even his body, dried and white as chalk, with little sparse hairs, as though somebody had planted them. Despite my respect for Lenin's genius, it seemed unnatural to me, and above all anti-Materialist and anti-Leninist, this mystical gathering about Lenin's mortal remains.

Even if we had not been idle we still would have wished to see Leningrad, the city of the Revolution and the city of many beauties. I approached Zhdanov concerning this, and he graciously agreed. But I also detected a certain reserve. The meeting lasted barely ten minutes. Nevertheless, he did not fail to ask me what I thought of Dimitrov's statement in *Pravda,* on the occasion of his visit to Bucharest, in which he urged the co-ordination of economic plans and the creation of a customs union between Bulgaria and Rumania. I replied that I did not like the statement, for it treated Bulgarian-Rumanian relations in isolation and was premature. Neither was Zhdanov satisfied with the statement, though he did not bring out his reasons; they came out soon after and will be presented later at great length.

Somewhere at about the same time there arrived in Moscow a representative of Yugoslavia's foreign trade, Bogdan Crnobrnja, and inasmuch as he could not overcome some basic obstacles with the Soviet agencies, he importuned me to go with him on a visit to Mikoyan, the Minister of Foreign Trade.

Mikoyan received us coldly, betraying impatience. Among our requests was one that the Soviets deliver to us the railroad cars from their zones of occupation which they had already promised us—inasmuch as many of these cars had been taken out of Yugoslavia, and the Russians could not use them because their track gauge was broader than ours.

"And how do you mean that we give them to you— under what conditions, at what price?" Mikoyan asked coldly.

I replied, "That you give them to us as gifts!"

He replied curtly, "I am not in the business of giving gifts, but trade."

In vain, too, were the efforts Crnobrnja and I made to change the agreement on the sale of Soviet films, which was unfair and damaging to Yugoslavia. Excusing himself on the grounds that the other East European countries might consider it a precedent, Mikoyan refused even to take up the question. He was quite different, however, when the subject turned to Yugoslav copper. He offered to pay in any currency or in kind, in advance, and in any amounts.

Thus we got nowhere with him except to prolong

sterile and endless negotiations. It was obvious—the wheels of the Soviet machine had ground to a halt as far as Yugoslavia was concerned.

However, the trip to Leningrad brought some relief and refreshment.

Until my visit to Leningrad I would not have believed that anything could outdo the efforts of the natives of rebel regions and the Partisans of Yugoslavia in sacrifice and heroism. But Leningrad surpassed the reality of the Yugoslav revolution, if not in heroism then certainly in collective sacrifice. In that city of millions, cut off from the rear, without fuel or food, under the constant pounding of heavy artillery and planes, about three hundred thousand people died of hunger and cold during the winter of 1941-1942. Men were reduced to cannibalism, but there was no idea of surrendering. Yet that is only the general picture. Only after we came into contact with the realities—with concrete cases of sacrifice and heroism and with the living men who were involved or were their witness—did we feel the grandeur of the epic of Leningrad and the strength of what human beings—the Russian people—are capable of when the foundations of their spiritual, political, and general existence are endangered.

Our encounter with Leningrad's officials added human warmth to our admiration. They were all, to a man, simple, educated, hard-working people who had taken on their shoulders and still bore in their hearts the tragic greatness of the city. But they lived lonely lives and were glad to meet men from another clime and culture. We got along with them easily and quickly—as men who had experienced a similar fate. Though it never occurred to us to complain about the Soviet leaders, still we observed that these men approached the life of their city and citizens—that most cultured and most industrialized center in the vast Russian land—in a simpler and more human way than was the case in Moscow.

It seemed to me that I could very quickly arrive at a common political language with these people simply by employing the language of humanity. Indeed, I was not surprised to hear two years later that these people, too, had failed to escape the totalitarian millstone just because they dared also to be men.

In this radiant, yet sad, Leningrad episode of ours there was also an unpleasant blot—our escort, Lesakov. Even at that time one encountered officials in the Soviet Union who had emerged from the lower strata of the working masses. One could tell—by his inadequate literacy and rusticity—that Lesakov had recently been a worker. These deficiencies would not have been vices had he not tried to conceal them and had he not made a conspicuous display of pretensions beyond his capacities. In actual fact, he had not made his way up by dint of his own forces and abilities, but he had been dragged to the top and planted in the apparatus of the Central Committee, in which he was charged with Yugoslav affairs. He was a cross between an intelligence agent and a Party official. Cast in the role of the Party man and "Partyness," he collected in crude fashion information about the Yugoslav Party and its leaders.

Slight as he was, with a knotty face and short yellow

teeth, a tie that hung crookedly and a shirt that kept spilling out of his pants, always afraid he might look "uncultured," Lesakov would have been pleasant as an ordinary workingman had he not been charged with such a great duty and hence kept provoking us—mostly me—into unpleasant discussions. He boasted of how "Comrade Zhdanov purged all the Jews from the apparatus of the Central Committee!"—and yet he simultaneously lauded the Hungarian Politburo, which at that time consisted almost entirely of Jewish émigrés, which must have suggested to me the idea that, despite its covert anti-Semitism, the Soviet Government found it convenient to have Jews at the top in Hungary because they were rootless and thus all the more dependent upon its will.

I had already heard and observed that when they want to get rid of someone in the Soviet Union but lack convincing reasons for this, they usually spread some infamy about him through agents of the Secret Police. So it was that Lesakov told me "in confidence" that Marshal Zhukov had been ousted for looting jewelry in Berlin— "You know, Comrade Stalin cannot endure immorality!" —and about the Assistant Chief of the General Staff, General Antonov: "Imagine, he was exposed as being of Jewish origin!"

It was obvious, too, that Lesakov was, despite the limitations of his intelligence, well informed concerning affairs in the Yugoslav Central Committee and the methods of its work. "In no Party in Eastern Europe," said he, "is there such a closely watched foursome as yours."

He did not mention the names of that foursome, but I knew without asking that he was referring to Tito,

Kardelj, Ranković, and myself. And I asked myself and concluded: Is not that foursome also one of those "peanuts" in the eyes of the Soviet leadership?

7

After days of idleness, Koča Popović decided to return to our country, leaving Todorović in Moscow to attend the outcome, that is, to wait for the Soviet leadership to take pity and to resume talks. I would have gone off with Popović had not a message arrived from Belgrade announcing the arrival of Kardelj and Bakarić, and thus I had to join them in conversations with the Soviet Government concerning "the complications that had set in."

Kardelj and Bakarić arrived on Sunday, February 8, 1948. The Soviet Government had in fact invited Tito, but in Belgrade they made the excuse that Tito was not feeling well—even from this, one could see the mutual distrust—so Kardelj came in his stead. Invited simultaneously was a delegation from the Bulgarian Government, that is, the Central Committee, about which the ubiquitous Lesakov informed us, purposely underlining that the "top brass" had arrived from Bulgaria.

Prior to that, on January 29, *Pravda* had disavowed Dimitrov and disassociated itself from his "problematic and fantastic federations and confederations" and customs unions. This was an admonition and a foretaste of the tangible measures and stiffer course that the Soviet Government would undertake.

Kardelj and Bakarić were lodged in a villa near Moscow, and so I moved in with them. That same night, while Kardelj's wife was sleeping, and Kardelj was lying next to her, I sat down on the bed next to him and, as softly as I could, informed him of my impressions from my stay in Moscow and of my contacts with the Soviet leaders. They boiled down to the conclusion that we could not count on any serious help but had to rely on our resources, for the Soviet Government was carrying on its own policy of subordination, trying to force Yugoslavia down to the level of the occupied East European countries.

Kardelj told me, then or just after his arrival, that the direct cause of the dispute with Moscow was the agreement between the Yugoslav and Albanian Governments regarding the entry of two Yugoslav divisions into Albania. The divisions were already being formed, while a regiment of the Yugoslav fighter Air Force was already in Albania when Moscow vigorously protested, refusing to accept as a reason that the Yugoslav divisions were needed to defend Albania from possible attack by the Greek "monarcho-fascists." In his dispatch to Belgrade, Molotov threatened an open breach.

The day after Kardelj's arrival, while promenading in the park under the surveillance of Soviet agents on whose faces we read fury at our having a conference that they could not overhear, Kardelj and I continued our conversation, in Bakarić's presence. It was broader and more consequential in its analyses, and, despite insignificant differences in our conclusions, completely unanimous. As usual, I was the more severe and peremptory one.

No one informed us of anything and there was not a

sign from the Soviet side until the next evening, February 10, when they picked us up in a car at nine o'clock and drove us to the Kremlin, to Stalin's office. There we waited fifteen minutes or so for the Bulgars—Dimitrov, Kolarov, and Kostov—and as soon as they arrived, we were all immediately taken in to Stalin. We were seated so that to the right of Stalin, who was at the head, sat the Soviet representatives—Molotov, Zhdanov, Malenkov, Suslov, Zorin; to the left were the Bulgars—Kolarov, Dimitrov, Kostov; then the Yugoslav representatives— Kardelj, myself, Bakarić.

At the time, I submitted a written report of that meeting to the Yugoslav Central Committee, but inasmuch as I am not able to get at it today, I shall rely on my memory and on what has already been published about the meeting.

The first to be recognized was Molotov, who, with customary terseness, brought out that serious differences had appeared between the Soviet Government on the one hand and the Yugoslav and Bulgarian Governments on the other hand, which was "impermissible from both the Party and the political point of view."

As examples of these differences he cited the fact that Yugoslavia and Bulgaria had signed a treaty of alliance not only without the knowledge of, but contrary to, the views of the Soviet Government, which held that Bulgaria should not sign any political treaties before signing a peace treaty.

Molotov wished to dwell rather longer on Dimitrov's statement in Bucharest concerning the creation of an East European Federation, in which Greece was included, and a customs union and co-ordination of economic plans between Rumania and Bulgaria. However, Stalin cut him short. "Comrade Dimitrov gets too carried away at press conferences—doesn't watch what he's saying. And everything he says, that Tito says, is taken abroad to be with our knowledge. For example, the Poles have been visiting here. I ask them: What do you think of Dimitrov's statement? They say: A good thing. And I tell them that it isn't a good thing. Then they reply that they, too, think it isn't a good thing—if that is the opinion of the Soviet Government. For they thought that Dimitrov had issued that statement with the knowledge and concurrence of the Soviet Government, and so they approved of it. Dimitrov later tried to amend that statement through the Bulgarian telegraph agency, but he didn't help matters at all. Moreover, he cited how Austria-Hungary had in its day obstructed a customs union between Bulgaria and Serbia, which naturally prompts the conclusion: the Germans were in the way earlier, now it is the Russians. There, that's what is going on!"

Molotov continued that the Bulgarian Government was going ahead with establishing a federation with Rumania without even consulting the Soviet Government about this.

Dimitrov attempted to temper the affair, emphasizing that he had spoken of federation only in general terms.

But Stalin interrupted him: "No, you agreed on a customs union, on the co-ordination of economic plans."

Molotov followed up Stalin: ". . . and what is a customs union and co-ordination of economics but the creation of a state?"

At that moment the substance of the meeting came drastically into view, though no one expressed it, namely: no relations among the "people's democracies" were permissible beyond the interests and without the approval of the Soviet Government. It became evident that to the Soviet leaders, with their great-power mentality (which found expression in the concept of the Soviet Union as "the leading force of socialism"), and especially with their cognizance that the Red Army had liberated Rumania and Bulgaria, Dimitrov's statements and Yugoslavia's lack of discipline and willfulness were not only heresy but the denial of the Soviet Union's "sacred" rights.

Dimitrov tried to explain, to justify himself, but Stalin kept interrupting without letting him finish. This was now the real Stalin. His wit now turned into malicious crudity, and his exclusiveness into intolerance. Still he kept restraining himself and succeeded in not going berserk. Without losing even for a moment his feel for the actual state of affairs, he upbraided the Bulgars and bitterly reproached them, for he knew they would submit to him, but in fact he had his sights fixed on the Yugoslavs—according to the folk saying: She scolds her daughter in order to reproach her daughter-in-law.

Supported by Kardelj, Dimitrov pointed out that Yugoslavia and Bulgaria had not announced a signed treaty at Bled but only a statement that an agreement had been reached leading to a treaty.

"Yes, but you didn't consult with us!" Stalin shouted. "We learn about your doings in the newspapers! You chatter like women from the housetops whatever occurs to you, and then the newspapermen grab hold of it!"

Dimitrov continued, obliquely justifying his position on the customs union with Rumania, "Bulgaria is in such economic difficulties that without co-operation with other countries it cannot develop. As far as my statement at the press conference is concerned, it is true that I was carried away."

Stalin interrupted him, "You wanted to shine with originality! It was completely wrong, for such a federation is inconceivable. What historic ties are there between Bulgaria and Rumania? None! And we need not speak of Bulgaria and, let us say, Hungary or Poland."

Dimitrov remonstrated, "There are essentially no differences between the foreign policies of Bulgaria and the Soviet Union."

Stalin, decidedly and firmly: "There are serious differences. Why hide it? It was Lenin's practice always to recognize errors and to remove them as quickly as possible."

Dimitrov, placatingly, almost submissively: "True, we erred. But through errors we are learning our way in foreign politics."

Stalin, harshly and tauntingly: "Learning! You have been in politics fifty years—and now you are correcting errors! Your trouble is not errors, but a stand different from ours."

I glanced sidelong at Dimitrov. His ears were red, and big red blotches cropped up on his face covering his spots of eczema. His sparse hair straggled and hung in lifeless strands over his wrinkled neck. I felt sorry for him. The lion of the Leipzig Trials, who had defied Göring and fascism from his trap at the time of their greatest ascendancy, now looked dejected and dispirited.

Stalin went on: "A customs union, a federation between Rumania and Bulgaria—this is nonsense! A federation between Yugoslavia, Bulgaria, and Albania is another matter. Here there exist historic and other ties. This is the federation that should be created, and the sooner, the better. Yes, the sooner, the better—right away, if possible, tomorrow! Yes, tomorrow, if possible! Agree on it immediately."

Someone, I think it was Kardelj, observed that a Yugoslav-Albanian federation was already in the making.

But Stalin stressed, "No, first a federation between Bulgaria and Yugoslavia, and then both with Albania."

And then he added, "We think that a federation ought to be formed between Rumania and Hungary, and also Poland and Czechoslovakia."

The discussion calmed down for a moment.

Stalin did not develop this question of federation further. He did repeat later, in the form of a directive, that a federation between Yugoslavia, Bulgaria, and Albania should immediately be formed. But from his stated position and from vague allusions by Soviet diplomats at the time, it seemed that the Soviet leaders were also toying with the thought of reorganizing the Soviet Union by joining to it the "people's democracies"—the Ukraine with Hungary and Rumania, and Byelorussia with Poland and Czechoslovakia, while the Balkan states were to be joined with Russia! However obscure and hypothetical all these plans may have been, one thing is certain: Stalin sought solutions and forms for the East European countries that would solidify and secure Moscow's domination and hegemony for a long time to come.

Just as it seemed that the question of a customs union, that is, the Bulgarian-Rumanian agreement, had been settled, old Kolarov, as though recalling something important, began to expound. "I cannot see where Comrade Dimitrov erred, for we previously sent a draft of the treaty with Rumania to the Soviet Government and the Soviet Government made no comment regarding the customs union except with regard to the definition of the aggressor."

Stalin turned to Molotov: "Had they sent us a draft of the treaty?"

Molotov, without getting confused, but also not without acrimony: "Well, yes."

Stalin, with angry resignation: "We, too, commit stupidities."

Dimitrov latched on to this new fact. "This was precisely the reason for my statement. The draft had been sent to Moscow. I did not suppose that you could have anything against it."

But Stalin remained unyielding. "Nonsense. You rushed headlong like a Komsomol youth. You wanted to astound the world, as though you were still Secretary of the Comintern. You and the Yugoslavs do not let anyone know what you are doing, but we have to find out everything on the street. You place us before the accomplished fact!"

Kostov, who was in charge of Bulgaria's economic affairs at the time, wished to say something too. "It is hard to be a small and underdeveloped country. . . . I would like to raise some economic questions."

But Stalin cut him short, directing him to the competent ministries and pointing out that this was a meeting to discuss the differences in foreign policy of three governments and parties.

Finally Kardelj was recognized. He was red and, what was a sign of agitation with him, he drew his head down between his shoulders and made pauses in his sentences where they did not belong. He pointed out that the treaty between Yugoslavia and Bulgaria, signed at Bled, had been previously submitted to the Soviet Government, but that the Soviet Government had made no comment other than with respect to its duration—instead of "for all time," it suggested "twenty years."

Stalin kept glancing silently and not without reproach at Molotov, who hung his head and with clenched lips in fact confirmed Kardelj's claim.

"Except for that comment, which we adopted," Kardelj corroborated, "there were no differences. . . ."

Stalin interrupted him, no less angrily though less offensively than with Dimitrov. "Nonsense! There are differences, and grave ones! What do you say about Albania? You did not consult us at all regarding the entry of your army into Albania."

Kardelj stressed that there existed the consent of the Albanian Government for that.

Stalin shouted, "This could lead to serious international complications. Albania is an independent state!

What do you think? Justification or no justification, the fact remains that you did not consult us about the sending of two divisions into Albania."

Kardelj explained that all that had not yet been final and added that he did not remember a single foreign problem but that the Yugoslav Government did not consult with the Soviet.

"It's not so!" Stalin cried. "You don't consult at all. That is not your mistake, but your policy—yes, your policy!"

Cut off, Kardelj fell silent and did not press his view.

Molotov took up a piece of paper and read a passage from the Yugoslav-Bulgarian treaty: that Bulgaria and Yugoslavia would "work in the spirit of the United Nations and support all action directed at the preservation of peace and against all hotbeds of aggression."

"What is the meaning of this?" Molotov asked.

Dimitrov explained that these words signified solidarity with the United Nations in the struggle against hotbeds of aggression.

Stalin broke in: "No, this is preventive war—the commonest Komsomol stunt; a tawdry phrase which only brings grist to the enemy mill."

Molotov returned to the Bulgarian-Rumanian customs union, underscoring that this was the beginning of a merger between the two states.

Stalin broke in with the observation that customs unions are generally unrealistic. Since the discussion had again subsided somewhat, Kardelj observed that some customs unions had shown themselves not to be so bad in practice.

"For example?" Stalin asked.

"Well, for example, Benelux," Kardelj said cautiously. "Here Belgium, Holland, and Luxembourg joined together."

Stalin: "No, Holland didn't. Only Belgium and Luxembourg. That's nothing, insignificant."

Kardelj: "No, Holland is included too."

Stalin stubbornly: "No, Holland is not."

Stalin looked at Molotov, at Zorin, at the rest. I had the desire to explain to him that the syllable *ne* in the name Benelux came from the Netherlands, that is, the original designation for Holland, but since everyone kept still, I did too, and so it remained that Holland was not in Benelux.

Stalin returned to the co-ordination of economic plans between Rumania and Bulgaria. "That is senseless, for instead of co-operation there would soon be a quarrel. The unification of Bulgaria and Yugoslavia is another matter—there are similarities here, ancient aspirations."

Kardelj pointed out that at Bled it had also been decided to work gradually toward federation between Bulgaria and Yugoslavia, but Stalin broke in by being more precise: "No, but immediately—by tomorrow! First Bulgaria and Yugoslavia ought to unite, and then let Albania join them later."

Stalin then turned to the uprising in Greece: "The uprising in Greece has to fold up." (He used for this the word *svernut'*, which means literally *to roll up*.) "Do you believe"—he turned to Kardelj—"in the success of the uprising in Greece?"

Kardelj replied, "If foreign intervention does not grow

and if serious political and military errors are not made."

Stalin went on, without paying attention to Kardelj's opinion: "If, if! No, they have no prospect of success at all. What do you think, that Great Britain and the United States—the United States, the most powerful state in the world—will permit you to break their line of communication in the Mediterranean Sea! Nonsense. And we have no navy. The uprising in Greece must be stopped, and as quickly as possible."

Someone mentioned the recent successes of the Chinese Communists. But Stalin remained adamant: "Yes, the Chinese comrades have succeeded, but in Greece there is an entirely different situation. The United States is directly engaged there—the strongest state in the world. China is a different case, relations in the Far East are different. True, we, too, can make a mistake! Here, when the war with Japan ended, we invited the Chinese comrades to reach an agreement as to how a modus vivendi with Chiang Kai-shek might be found. They agreed with us in word, but in deed they did it their own way when they got home: they mustered their forces and struck. It has been shown that they were right, and not we. But Greece is a different case—we should not hesitate, but let us put an end to the Greek uprising."

Not even today am I clear on the motives that caused Stalin to be against the uprising in Greece. Perhaps he reasoned that the creation in the Balkans of still another Communist state—Greece—in circumstances when not even the others were reliable and subservient, could hardly have been in his interest, not to speak of possible international complications, which were assuming an increasingly threatening shape and could, if not drag him into war, then endanger his already-won positions.

As far as the pacification of the Chinese revolution was concerned, here he was undoubtedly led by opportunism in his foreign policy, nor can it be excluded that he anticipated future danger to his own work and to his own empire from the new Communist great power, especially since there were no prospects of subordinating it internally. At any rate, he knew that every revolution, simply by virtue of being new, also becomes a separate epicenter and shapes its own government and state, and this was what he feared in the Chinese case, all the more since the phenomenon was involved that was as significant and as momentous as the October Revolution.

The discussion began to flag in tempo, and Dimitrov mentioned the development of further economic relations with the USSR, but Stalin cut him off again: "We shall speak of this with the joint Bulgarian-Yugoslav Government."

To Kostov's complaint concerning the injustice of an agreement on technical aid, Stalin replied that he submit a note—"*zapisochka*"—to Molotov.

Kardelj asked what stand should be taken concerning the demand of the Italian Government that Somalia be placed under its trusteeship. Yugoslavia was not inclined to support that demand, but Stalin held the opposite view and he asked Molotov if a reply had been made to that effect. He motivated his stand thus: "Once kings, when they could not agree over the booty, used to give disputed territories to their weakest vassal so they could snatch them from him later at some opportune moment."

Stalin did not forget, somewhere before the close of the meeting, to cloak the reality—his demands and orders—with Lenin and Leninism. He declared, "We too, Lenin's disciples, often had differences with Lenin himself, and even quarreled over some thing, but later we would talk it all out, establish our positions and—we would go forward."

The meeting had lasted about two hours.

This time Stalin did not invite us to dinner in his home. I must confess that I felt a sadness and an emptiness because of this, so great was my own human, sentimental fondness for him still.

I felt a cold void and bitterness. In the car I tried to express to Kardelj my indignation over the meeting, but, being crushed, he gave me a sign to be still.

That does not mean that we did not agree, but we reacted in different ways.

How great was Kardelj's confusion was most evident the next day, when they took him to the Kremlin to sign —without explanation or ceremony—with Molotov a treaty on consultation between the USSR and Yugoslavia, and he put his signature in the wrong place, so that he had to sign over again.

The same day, according to an agreement made in Stalin's anteroom, we went to Dimitrov's for lunch—to agree on a federation. We did it mechanically—the remnant of discipline and the authority of the Soviet Government. The conversation over this was short and listless on both sides; we agreed that we would get in touch as soon as all had arrived in Sofia and Belgrade.

To be sure, all this came to nought, for a month later Molotov and Stalin began to attack the Yugoslav leadership in their letters, finding in this the support of the Bulgarian Central Committee. The federation with Bulgaria turned out to be a snare—to crack the unity of the Yugoslav Communists—a snare into which no idealist wished to place his neck any longer. Although on the surface all was quieted, and it appeared that we were united, the protagonists were taking exaggerated positions. This was the prelude to what was to come later, the open division between the Soviet Union and Yugoslavia, which occurred in June of 1948.

There has remained in my memory of that meeting with the Bulgarian delegation the remembrance of Kostov's amenity, almost tenderness toward us. This was all the more unusual inasmuch as in high Yugoslav Communist circles he was considered an opponent of Yugoslavia, and by the same token a Soviet man. Yet he was also for Bulgarian independence, and therefore looked on the Yugoslavs with displeasure in the belief that they were the chief henchmen of the Soviets, and even inclined to place Bulgaria and its Communist Party under themselves. Kostov was later shot on the false charge that he was in the service of Yugoslavia, while the Yugoslav press kept attacking him, so to speak, to the last day—such was the distrust and misunderstanding under Stalin's shadow.

It was on that occasion that Dimitrov said what he did about the atom bomb, and then, as though in passing, while accompanying us out of his villa, "What is involved here is not criticism of my statement but something else."

Dimitrov certainly knew as much as we did. But he did not have the forces, and perhaps he himself lacked the strength of the Yugoslav leaders.

I did not fear that anything could happen to us in Moscow; after all, we were the representatives of a foreign independent state. And yet there frequently came before my eyes the vision of Bosnian forests, in whose depths we found refuges during the most violent German offensives and at whose clear cold springs we always found rest and comfort. I even told Kardelj, or someone else, thus laying myself open to reproof for exaggerating, "Just as long as we get to our hills and forests as soon as possible!"

We left three or four days later. They drove us to the airport at Vnukovo at dawn and stuck us on the plane without any honors. As we flew, I felt more and more the happiness of a child, though also a serious, stern joy, and kept thinking less and less of Stalin's story concerning General Sikorski's fate.

Was I that same person who four years earlier had sped to the Soviet Union devoted and candid in all his being?

Once again a dream was snuffed out on contact with reality.

Could this mean that a new one might sprout?

Conclusion

MANY PERSONS, among them Trotsky, of course, stress Stalin's criminal, bloodthirsty passions. I have no intention of either denying or confirming them, since the facts are not that well known to me. Recently it was made public in Moscow that he had probably killed the Leningrad Secretary, Kirov, in order to gain a pretext for settling accounts with the intra-Party opposition. He probably had a hand in Gorky's death; that death was depicted too prominently in his propaganda as the work of the opposition. Trotsky even suspects that he killed Lenin, with the excuse that he was shortening his misery. It is claimed that he killed his own wife, or in any case, through his harshness, he caused her to kill herself. The romantic legend spread by Stalin's agents, and which I, too, had heard, is truly too naïve—that she was poisoned while tasting food before her good husband.

Every crime was possible to Stalin, for there was not one he had not committed. Whatever standards we use to take his measure, in any event—let us hope for all time to come—to him will fall the glory of being the greatest criminal in history. For in him was joined the criminal senselessness of a Caligula with the refinement of a Borgia and the brutality of a Tsar Ivan the Terrible.

I was more interested, and am more interested, in how such a dark, cunning, and cruel individual could ever have led one of the greatest and most powerful states, not just for a day or a year, but for thirty years! Until precisely this is explained by Stalin's present critics—I mean his successors—they will only confirm that in good part they are only continuing his work and that they contain in their own make-up those same elements—the same ideas, patterns, and methods that propelled him. For in carrying out his undertakings not only did Stalin find it to his advantage to deal with an exhausted and desperate Russian postrevolutionary society, but it is also true that certain strata of that society, to be more exact, the ruling political bureaucracy of the Party, found use for just such a man—one who was reckless in his determination and extremely practical in his fanaticism. The ruling Party followed him doggedly and obediently—and he truly led it from victory to victory, until, carried away by power, he began to sin against it as well. Today this is all it reproaches him for, passing in silence over his many greater and certainly no less brutal crimes against the "class enemy"—the peasantry and the intelligentsia, and also the left and right wings within the Party and outside of it. And as long as that Party fails to break, both in its theory and especially in its practice, with everything that comprised the very originality and essence of Stalin and of Stalinism, namely, with the ideological unitarianism and so-called monolithic structure of the Party, it will be a bad but reliable sign that it has not emerged from under Stalin's shadow. Thus the present joy over the liquidation of the so-called anti-Party group of Molotov, despite all the odiousness of his personality and the depravity of his views, seems to me to be shallow and premature. For the essence of the problem is not whether this group is better than that, but that they should exist at all—and whether, at least as a beginning, the ideological and political monopoly of a single group in the USSR shall be ended. Stalin's dark presence continues to hover and—assuming that there will not be a war—one can fear that it will hover over the Soviet Union for a relatively long time. Despite the curses against his name, Stalin still lives in the social and spiritual foundations of the Soviet society.

The references to Lenin in speeches and solemn declarations cannot change the substance. It is much easier to expose this or that crime of Stalin's than to conceal the fact that it was this man who "built socialism," who gave rise to the foundations of present Soviet society and of the Soviet empire. All this bespeaks the fact that Soviet society, despite its gigantic technical achievements, and perhaps largely because of them, has barely begun to change, that it is still imprisoned in its own, Stalinist, dogmatic framework.

Despite this criticism, the hopes do not seem entirely baseless that in the foreseeable future new ideas and phenomena may appear which, though they may not shake Khrushchev's "monolithism," will at least cast light on its contradictions and on its essence. At the moment the conditions for more substantial changes do not exist. Those who govern are still themselves too poor to find dogmatism and monopoly of rule a hindrance or superfluous, while the Soviet economy can still exist enclosed in its own empire and can absorb the losses caused by its separation from the world market.

To be sure, much that is human assumes proportions

and values according to the corner from which it is viewed.

So it is with Stalin.

If we assume the viewpoint of humanity and freedom, history does not know a despot as brutal and as cynical as Stalin was. He was methodical, all-embracing, and total as a criminal. He was one of those rare terrible dogmatists capable of destroying nine tenths of the human race to "make happy" the one tenth.

However, if we wish to determine what Stalin really meant in the history of Communism, then he must for the present be regarded as being, next to Lenin, the most grandiose figure. He did not substantially develop the ideas of Communism, but he championed them and brought them to realization in a society and a state. He did not construct an ideal society—something of the sort is not even possible in the very nature of humans and human society, but he transformed backward Russia into an industrial power and an empire that is ever more resolutely and implacably aspiring to world mastery.

Viewed from the standpoint of success and political adroitness, Stalin is hardly surpassed by any statesman of his time.

I am, of course, far from thinking that success in political struggles is the only value. It especially does not occur to me to identify politics with amorality, though I do not deny that, by the very fact that politics involve a struggle for the survival of given human communities, they are thereby marked by a disregard for moral norms. For me great politicians and great statesmen are those who can join ideas and realities, those who can go forward steadfastly toward their aims while at the same time adhering to the basic moral values.

All in all, Stalin was a monster who, while adhering to abstract, absolute, and fundamentally utopian ideas, in practice recognized, and could recognize, only success—violence, physical and spiritual extermination.

However, let us not be unjust toward Stalin! What he wished to accomplish, and even that which he did accomplish, could not be accomplished in any other way. The forces that swept him forward and that he led, with their absolute ideals, could have no other kind of leader but him, given that level of Russian and world relations, nor could they have been served by different methods. The creator of a closed social system, he was at the same time its instrument and, in changed circumstances and all too late, he became its victim. Unsurpassed in violence and crime, Stalin was no less the leader and organizer of a certain social system. Today he rates very low, pilloried for his "errors," through which the leaders of that same system intend to redeem both the system and themselves.

And yet, despite the fact that it was carried out in an inappropriate operetta style, Stalin's dethronement proves that the truth will out even if only after those who fought for it have perished. The human conscience is implacable and indestructible.

Unfortunately, even now, after the so-called de-Stalinization, the same conclusion can be reached as before: Those who wish to live and to survive in a world different from the one Stalin created and which in essence and in full force still exists must fight.

SELECTED BIOGRAPHICAL NOTES

(Prepared by the Publisher)

GEORGI FEDOROVICH ALEKSANDROV (1908-   )

Leading Soviet philosopher and Communist Party member since 1928. He worked in the Agitation and Propaganda Section (Agit-prop) of the Central Committee from 1934 and was its head from 1939 to 1947. His book *History of Western European Philosophy in the Nineteenth Century*, published in 1944, was officially attacked by Zhdanov for presenting Marxism as a part of the Western philosophical tradition. In 1950 he was official commentator on the philosophical implications of Stalin's articles on linguistics. He served as Minister of Culture in 1954-1955, after which he joined the Institute of Philosophy of the Byelorussian Academy of Sciences in Minsk.

VLADIMIR BAKARIĆ (1912-   )

Croatian who joined the Communist underground in 1933 as a student and was sentenced in 1934 to three years in prison. In 1941 he joined the Partisans. After the war he became Premier of Croatia. In 1946 he was a member of the Yugoslav delegation to the Peace Conference in Paris. He is the ranking Communist leader in Croatia.

LAVRENTY PAVLOVICH BERIA (1899-1953)

Georgian Communist who made a career in the Soviet Secret Police—the Cheka, GPU, and NKVD. As Commissar for Internal Affairs from 1938 to 1948 and Deputy Prime Minister in charge of security from 1941 to 1953, he ended the Great Purge by liquidating his predecessor, N. I. Yezhov, and many other officials and also directed the reign of terror, not only in the Soviet Union but in the satellite states, that marked Stalin's last years. He was purged in the power struggle following Stalin's death.

SEMËN MIKHAILOVICH BUDËNNY (1883-   )

Marshal of the Soviet Union, from 1935. He was active in the Revolution of 1917. From 1939 he has been a member of the

Central Committee of the Communist Party, and in 1940 was First Vice-Commissar of Defense.

NIKOLAI IVANOVICH BUKHARIN (1888-1938)

Leading Bolshevik theorist and member of the Politburo from 1918 to 1929 who supported Stalin against Trotsky but was himself stripped of power by Stalin as leader of the Right Opposition and executed during the Great Purge. Many of his ideas have found expression in post-Stalin revisionism, especially in Poland, Hungary, and East Germany.

NIKOLAI ALEXANDROVICH BULGANIN (1895-   )

Soviet politician. He joined the Communist Party in 1917, and was a member of the Supreme Soviet from 1937 to 1958. From 1941 to 1944 he was a member of the Military Council, and the following year served on the State Defense Committee. Other posts he has held have been: Deputy People's Commissar of Defense (1944-1947), Minister of Defense (1947-1949 and 1953-1955), Chairman of the Council of Ministers (1955-1958), member of the Politburo (1948-1952), member of the Presidium (1952-1958), and Prime Minister (1955-1958).

VLKO ČHERVENKOV (1900-   )

Bulgarian Communist leader who joined the Party in 1919. He was forced to flee Bulgaria for the USSR in 1925 with his wife, Dimitrov's sister, for his complicity in the infamous bombing in the Sofia Cathedral. He completed his studies at the Lenin Party School in the USSR and joined the Agitation and Propaganda Section of the Communist International. In 1937, during the Great Purge, he was made director of the Lenin School, which post he held until the school was closed in 1941. During the Second World War he managed the Soviet radio station Khristo Botev, which broadcast to Bulgaria. On September 9, 1944, he returned to Bulgaria to take over the Secretariat of the Communist Party. In January 1950 he succeeded Kolarov as Premier. In November of the same year he became Secretary General of the Party but gave up the post after Stalin's death. He served as Minister of Culture and was eventually reinstated in the Politburo.

BOGDAN CRNOBRNJA (1916-   )

Yugoslav teacher who joined the Partisans during the Second World War. After the liberation, he served as Deputy Minister of

Foreign Trade and of Foreign Affairs. Since 1955 he has been Yugoslav Ambassador to India.

PEKO DAPČEVIĆ (1913-   )

Communist Yugoslav general. He joined the Party in 1933 as a student at the University of Belgrade. His first military experience came in 1936 as a company commander in the International Brigade during the Spanish Civil War. With the invasion of Yugoslavia in 1941, he led the Partisan uprising in his native Montenegro and thereafter rose rapidly to the Supreme Headquarters of the Army of People's Liberation. In 1945 he was awarded the medal of People's Hero. The following year he commanded the Yugoslav Fourth Army in the Yugoslav zone of Venezia-Guilia, the hinterland of Trieste, and was then assigned to direct the guerrilla action in Northern Greece. From 1953 he served as Chief of the Yugoslav General Staff, but was demoted as a result of being indirectly implicated in the Djilas affair; it was his actress wife, Milena Vrajak, whom Djilas defended against the "New Class."

GEORGI DIMITROV (1882-1949)

Bulgarian Communist leader who was one of the organizers of the Bulgarian Communist Party in 1909. After a career as underground activist and union organizer in Bulgaria, he was released from prison through Russian intervention in 1921 and for the next two decades served in the Comintern. He was General Secretary of the Communist International in Moscow for nine years, and was the author of the Popular Front policy of the thirties. He gained world-wide prominence as a result of his trial, and acquittal, in Berlin in 1933 for complicity in the Reichstag fire. After the Second World War he gave up his Soviet citizenship and returned to Bulgaria to assume leadership of the Communists there and to carry out the Communization of that country. He became Premier in 1946.

MAXIM GORKY (1868-1936)

Russia's leading revolutionary novelist. His works—notably *Mother, The Artamonov Business,* and *Klim Samgin*—were a condemnation of capitalist society. Though he gave considerable financial support to the Bolsheviks, he opposed their seizure of power and lived in exile from 1921 to 1928. Upon his return, he headed the Writers' Union and was declared founder of the school of Socialist Realism. A close friend of Stalin's, he became a leading apologist for the Soviet regime. He died in allegedly mysterious circumstances in 1936. Official blame for his death was placed on the "Anti-Soviet Bloc of Rightists" and the Trotskyites during the Bukharin show trial of 1938. Since then, Stalin himself has been accused of complicity in his death.

ANDRIJA HEBRANG (1899-1948)

Yugoslav Communist leader from Croatia. He spent twelve years in prison before the Second World War for his activities in the trade-union movement. Upon his release he became Secretary of the Croatian Communist Party. He was a leader of the National Liberation Movement from the start, in 1941, and held high offices after the war, among them Minister of Industry, member of the Presidium of both the Yugoslav and Croatian Constituent Assemblies, and Chairman of the Federal Planning Commission. In 1946 the Party's Central Committee investigated his past and found him guilty of cowardice during the war and of collaboration with the Croatian Fascist Ustaši. He was also declared a "fractionist" and relieved of his posts. In 1948 he was arrested, allegedly while trying to escape to Rumania. He committed suicide while awaiting trial. Some sources claim he was murdered in jail.

ENVER HOXHA (1908-   )

Leading Albanian Communist leader. He was educated in France and Belgium and taught French in Albanian schools. He was a founder of the Albanian Communist Party in 1941 and of the Albanian National Liberation Movement in 1942. In 1943 he became Secretary General of the Albanian Communist Party, which post he held until 1954, when it was abolished. He has since served as First Secretary of the Party's Central Committee. In 1946 he was Premier, Foreign Minister, Defense Minister, and Commander in Chief of Albania's armed forces.

ARSO JOVANOVIĆ

Professional prewar Yugoslav army officer from Montenegro. He joined the Partisans and organized the People's Liberation Army, of which he was Chief of the General Staff until the end of 1946, when he was replaced by Koča Popović. He was openly on the side of the Soviet Union in the Tito-Cominform break in 1948. He was shot by border guards while trying to escape to Rumania.

LAZAR MOISEEVICH KAGANOVICH (1893-   )

Communist of humble Jewish origin who was a Party organization man. He rose to power as one of Stalin's chief henchmen. During the Second World War he was a member of the State Defense Committee and subsequently held high posts in the Caucasus and the Ukraine. His influence declined in Stalin's last years, perhaps in part because of the anti-Semitic campaign. After Stalin's death he became prominent once again, but was divested of all power in 1957 as a member of the "anti-Party group."

EDVARD KARDELJ (1907-   )

Yugoslav Communist leader generally regarded as second to Tito. A Slovenian schoolteacher, he joined the Party in 1928. He was jailed for two years in 1931. From 1934 to 1937 he studied in the Comintern's Lenin School in Moscow and served as a professor there. He collaborated with Tito in the reorganization of the Yugoslav Communist Party before the war, and became a member of its Politburo in 1940. During the war he served in the Partisan Supreme Command and became Vice-Premier of the Provisional Government founded in 1943. He retained this post when the Government was constitutionally established in 1945. Since 1951 he has also served as Foreign Minister and as president of the commission in charge of the reorganization of the Government. He is regarded as a top ideologist in the Yugoslav Communist Party.

NIKITA SERGEEVICH KHRUSHCHEV (1894-   )

Chairman of the Soviet Council of Ministers and First Secretary of the Central Committee of the Communist Party of the Soviet Union. A locksmith by trade, he rose through the ranks of the Communist Party especially through his activities in the Ukraine. Following the Civil War, in which he served as a political commissar of a Partisan detachment, he was sent to the Workers' School at Kharkov University. Thereafter he ascended the ladder of Party posts up to the Politburo (candidate member in 1935) and Central Committee. In 1938 he was put in charge of carrying out a purge in the Ukraine, and during the Second World War he served there in various army posts. After the war he was transferred from the Ukraine to Moscow, where he became a full member of the Party's Central Committee and Presidium in 1952. After Stalin's death, in 1953, he was elected First Secretary and eventually replaced Malenkov.

BORIS KIDRIČ (1919-1953)

Yugoslav Communist leader of Slovenian origin. He joined the

Party in 1928 and lived the life of a constantly hunted under-ground activist. He joined the Partisans in 1941, and became political commissar for Slovenia. In 1945 he was made Premier of Slovenia and continued a harsh program of establishing Communist hegemony there. In 1946 he was sent to Moscow to study the Soviet economy. From his return, in the fall of the same year, to his death, he was virtual director of the entire Yugoslav economy. His administration is associated with the ruthless collectivization of agriculture, abandoned after his death, and highly demanding production drives in industry. He was a member of the Politburo.

SERGEI MIRONOVICH KIROV (1886-1934)

Leading Bolshevik revolutionary and Politburo member in 1930. He at first supported Stalin in the latter's rise to power but opposed his personal rule after the Seventeenth Party Congress in 1934. His assassination in December 1934, probably at Stalin's behest, set off the Great Purge.

VASSIL KOLAROV (1877-1950)

Bulgarian Communist who succeeded Dimitrov as Premier in 1949. Like Dimitrov, he was a veteran of the Communist International and was its General Secretary in 1922. Following the Second World War, he left the USSR to return to Bulgaria, where he held the posts of Provisional President of the Bulgarian Republic (1946), Vice-President of the Council of Ministers and Foreign Minister (1949).

IVAN STEPANOVICH KONEV (1897-   )

Marshal of the Soviet Union. He distinguished himself during the Second World War, especially in the liberation of Kharkov (1943) and Kirovograd (1944). After the war he was Soviet representative on the Allied Control Commission in Vienna. From 1946 to 1955 he was Commander in Chief Land Forces, and from 1955 First Deputy Minister of Defense and Commander in Chief of the Warsaw Pact forces. He resigned from this post in 1960 on grounds of ill health. He was chairman of the special court that sentenced Beria in 1953.

TRAICHO KOSTOV (1897-1949)

Bulgarian Communist leader. He was a member of the Politburo and Deputy Prime Minister who, though an anti-Titoist, was associated with a "Bulgaria-first" outlook. Stripped of power in March 1949 and indicted in December of that year, he created a sensation by repudiating his confession at the trial. He was executed.

GEORGI MAXIMILIANOVICH MALENKOV (1902-   )

Soviet Communist Party leader who worked his way through the Party apparatus to become a member of the Central Committee, in 1939, where he was placed in charge of the administration of cadres. In 1941 he became a candidate member of the Politburo and served on the State Defense Committee throughout the Second World War. After the war he served as Secretary of the Central Committee and Deputy Prime Minister. He succeeded Stalin as Prime Minister in the era of "collective leadership" but was forced to step down after a public admission of failure in 1955. In 1957, as a member of the "anti-Party group," he was stripped of power.

DMITRI ZAKHAROVICH MANUILSKY (1883-   )

Soviet Communist Party official and diplomat. He joined the Party in 1903. As an underground activist, he experienced arrest and exile. After the Revolution of 1917 most of his posts were in his native Ukraine. However, he was even more active in the Communist International, serving as Secretary of the Presidium from 1928 to 1943. During the war he served as a political officer in the Red Army. He was also Foreign Minister for the Ukraine from 1945 to 1952, and head of the Ukrainian delegation to the United Nations in 1952.

ANASTAS IVANOVICH MIKOYAN (1895-   )

Armenian Communist who has been especially prominent as director of Soviet foreign trade and food industries. A candidate member of the Politburo in 1926, he became a full member in 1935. He has also been Deputy Prime Minister since 1937. After Stalin's death he consistently supported Khrushchev and has become one of the most influential leaders of the Soviet Communist Party. He is generally regarded as a "reasonable" Communist and achieved some popularity in the West, especially as a result of his American visit in 1958.

MITRA MITROVIĆ (1912-   )

Yugoslav Communist Party member since 1933. She was prominent in the Partisan ranks during the Second World War. After 1945 she served for several years as Minister of Education for Serbia. More recently she has risen to posts of federal rank in both the

Executive Council of the Government and the Central Committee of the Party.

VYACHESLAV MIKHAILOVICH MOLOTOV (1890-   )

A Bolshevik since 1906 and a specialist in Party organization. He ascended the ladder, largely as Stalin's lieutenant, until he was second in power only to Stalin. From 1926 he was a member of the Politburo and of the Presidium of the Executive Committee of the Comintern. He was Chairman of the Council of People's Commissars—that is, Prime Minister—throughout the thirties, and Deputy Chairman until 1957. He was best known to the world as Soviet Commissar (from 1946, Minister) for Foreign Affairs. In 1957 he was stripped of power as a member of the "anti-Party group" in association with Malenkov, Kaganovich, and others, and has since held relatively minor posts abroad.

BLAGOJE NEŠKOVIĆ (1907-   )

Serbian Communist who fought in the Spanish Civil War and joined Tito's Partisans in 1941. In 1945 he was Premier of Serbia. A member of the Central Committee of the Yugoslav Communist Party, he was accused of deviation in 1952 and stripped of his posts.

ANNA PAUKER (1893-   )

Rumanian Jew (nee Robinsohn) who was a founder of the Rumanian Communist Party in 1921. She married its head, Marcel Pauker. In 1924 both left Rumania for Moscow to work in Comintern headquarters. In 1936 she returned to Rumania, where she was arrested; her husband, in Moscow, fell in the Great Purge. Returned to Moscow in an exchange of prisoners, she became a member of the Executive Committee of the Comintern. During the Second World War she directed the Soviet radio station for broadcasts to Rumania and helped organize the Tudor Vladimirescu Division of Rumanian prisoners of war in the USSR. She returned to Rumania with the Red Army, and on November 7, 1947 became Foreign Minister and, soon after, Vice-Premier as well. In 1952 she fell from power, as a "deviationist."

MOŠA PIJADE (1889-195?)

Theoretician of the Yugoslav Communist Party. He was the oldest member in the Party when it was organized in 1920. Sentenced to twenty years in prison for spreading Communism in trade unions, he translated Marx's *Das Kapital* while serving his term in Sremska

Mitrovica Penitentiary, the same jail to which Djilas was later sentenced under the Tito regime. During the Second World War he and Djilas led the uprising in Montenegro, which gave rise to a ruthless civil war in that province. After the war he served as Vice-President of the Constituent Assembly and, later, of the Federal People's Assembly. In 1954, as a result of the Djilas affair, he became President of the Assembly. A member of the Yugoslav Communist Central Committee and Politburo, he was, until his death, in the inner circle around Tito.

KOČA POPOVIĆ (1908-   )

Scion of a prominent Belgrade family, Paris-trained lawyer, and poet. He joined the Yugoslav Communist Party in 1933 and fought in the Spanish Civil War. Upon his return he was arrested, but continued his underground activities after being released. In 1941 he joined the Partisans and rose to the highest military and Government echelons. He served as Chief of the General Staff from 1945 to 1953. Since 1946 he has held the post of Foreign Minister of Yugoslavia.

ALEKSANDAR-MARKO RANKOVIĆ (1909-   )

Yugoslav Communist Party leader, who joined the Serbian Youth Section of the Party in 1927. He spent five years in various prisons, where he got to know Tito and Pijade. In 1937, when Tito reorganized the Party, he was in the Politburo and has remained a top Communist ever since. After the liberation struggle, of which he was a leading organizer, he became best known as Minister of Interior and director of the Military and Secret Police. He and Kardelj are generally regarded as being next to Tito in power.

KONSTANTIN KONSTANTINOVICH ROKOSSOVSKY (1896-   )

A native Pole who joined the Red Army in 1919 and made a brilliant military career in the Soviet Union. He was one of the USSR's most outstanding generals during the Second World War. For his part in the defense of Moscow, Stalingrad, and Kursk, he was twice awarded the title of Hero of the Soviet Union, and became a Marshal in 1944. In 1949 he was officially transferred to the Polish Army and held the posts of Polish Minister of Defense, Commander in Chief, Deputy Prime Minister, and member of the Politburo of the Polish Communist Party. In November 1956 the

Gomulka regime had him transferred back to the Soviet Union, where he has since served as Deputy Minister of Defense.

PAVLE SAVIĆ (1909-   )

Paris-trained Yugoslav nuclear physicist and member of the Communist Party since 1939. He fought in the liberation struggle and was attached to Supreme Headquarters. In 1949 he received an award for his work with low temperatures.

RUDOLF SLANSKY (1901-1952)

Czech Communist leader. He became director of the Communist daily *Rudé Právo* in 1926. In 1928 he was elected to the Central Committee of the Party. He was a member of the Czechoslovak delegation to the last congress of the Comintern, in 1935. After Czechoslovakia's partition by Hitler, Slansky fled to the USSR, where he worked until 1944 in the Comintern. He returned to Czechoslovakia with the Red Army and became Secretary General of the reconstituted Czechoslovak Communist Party. He attended the Cominform meetings of 1947, 1948, and 1949. In September 1951 he was demoted from his leadership and, three months later, was arrested for "criminal activities." In 1952 he was hanged.

IVAN ŠUBAŠIĆ (1892-1955)

Croatian politician. He was Governor (*Ban*) of Croatia from August 1939, and went into exile during the war. On June 1, 1944 he was appointed Premier of the Yugoslav Royal Government-in-exile at the insistence of the Allies. He merged his cabinet with Tito's after the Tito-Šubašić Agreement concluded on the island of Vis. In the coalition Provisional Government, he served for a time as Foreign Minister.

MIKHAIL ANDREYEVICH SUSLOV (1902-   )

Communist Party leader in the USSR. He joined the Party in 1921, entered the Central Committee in 1941, and was a high-ranking political officer during the war. In 1946 he became head of the Agitation and Propaganda Section of the Central Committee, and in 1947 Secretary. In 1949-1950 he served as editor in chief of *Pravda*. His chief posts since then have been chairman of the Foreign Affairs Committee of the Soviet Union (1954) and member of the Central Committee's Presidium (1955). Generally regarded as a doctrinaire, he has nevertheless supported Khrushchev in defeating the "anti-Party group."

MIJALKO TODOROVIĆ (1913-   )

Yugoslav Communist leader. He began his Party career in the youth movement. He fought in the Partisan ranks during the Second World War. After the liberation he served in the Ministry of Defense, as Director of the Extraordinary Administration of Supply, Minister of Agriculture, and Chief of the Council for Agriculture and Forestry.

ALEKSANDR MIKHAILOVICH VASILEVSKY (1895-   )

Leading Soviet general and Chief of the Soviet General Staff at the time of the Battle of Stalingrad. He was made a Marshal in 1943, and was commander of the Byelorussian Front in 1945. Since then he has served as Minister of War.

NIKOLAI F. VATUTIN (1901-1944)

Soviet general. With Konev and Malinovsky, he distinguished himself in the liberation of the Ukraine from the German Army.

VELJKO VLAHOVIĆ (1914-   )

Montenegrin member of the Yugoslav Communist Party since 1935. He fought in the Spanish Civil War and was especially active in organizing the Communist Youth League of Yugoslavia. During the Second World War he directed the Free Yugoslavia radio station. He returned to Yugoslavia at the end of 1944 to serve as editor of the Communist daily, *Borba*, and as Deputy Foreign Minister. He has gained considerable reputation as a theoretician, especially since Djilas's fall.

NIKOLAI ALEKSEYEVICH VOZNESENSKY (1903-1950)

Leading Soviet economist. During the Great Purge, he rose rapidly to the post of Chairman of the State Planning Commission (Gosplan), which plans and co-ordinates the whole Soviet economy. He was also Deputy Prime Minister in 1939 and a member of the State Defense Committee during the war. Candidate member of the Politburo in 1941 and full member in 1948, he was stripped of all his posts in 1949 during Malenkov's campaign against Zhdanov's followers, and was arrested and shot on Stalin's orders.

SVETOZAR VUKMANOVIĆ-TEMPO (1912-   )

Montenegrin who joined Yugoslav Communist Youth in 1933 and became a Party member in 1935. His specialty in underground work was organizing clandestine presses. During the Second World War he served in Partisan Supreme Headquarters and was Tito's personal representative in Macedonia. In 1943 he was Chief Political Commissar in the People's Liberation Army. After the war he was active in the Federal Assembly and Central Planning and Central Economic Commissions. He is one of the closest collaborators of Tito.

KOČI XOXE (d.1948)

Albanian Communist leader who, thanks to Yugoslav backing, became the most powerful man in the Albanian Communist Party just after the Second World War, as Minister of the Interior and head of the Secret Police. At the time of the Tito-Cominform break, he was executed on charges of Trotskyite and Titoist activities.

ANDREI ALEKSANDROVICH ZHDANOV (1896-1948)

Secretary of the Soviet Communist Party Central Committee from 1935. He was a candidate member of the Politburo in 1934 and a full member in 1939. In charge of ideological affairs, he made Socialist Realism in the arts obligatory and directed the postwar campaign against Western cultural influences. During the Second World War he was a leader in the defense of Leningrad. He was prominent in the founding of the Cominform.

GEORGI KONSTANTINOVICH ZHUKOV (1894-   )

Marshal of the Soviet Union. He served in the Bolshevik forces in 1917. In 1941 he was Chief of Staff of the Red Army and conducted the defense of Moscow against the Germans. He was First Vice-Commissar of Defense in 1942, and the following year was promoted to Marshal.

VALERIAN ALEXANDROVICH ZORIN (1902-   )

Soviet diplomat. Among the posts he has held have been: Assistant General Secretary of the National Commissariat of Foreign Affairs (1941), Ambassador to Czechoslovakia (1945-1948), Deputy Minister of Foreign Affairs (1948), and Ambassador to the German Federal Republic (1956-1958). Since 1960 he has been Permanent Soviet Representative to the United Nations.

MIKHAIL MIKHAILOVICH ZOSHCHENKO (1895-1958)

Soviet author best known for his satirical works and his treatment of the bewildered "little man" in Soviet society. In 1946 Zhdanov made him a prime target in the Party campaign to impose its control over cultural life. He was expelled from the Writers' Union and lived in obscurity until his death.

INDEX

MILOVAN DJILAS was free on parole for fifteen months following his imprisonment for more than four years on charges of "slandering" and writing opinions "hostile to the people and the state of Yugoslavia." He was, up to the time of his expulsion from the Communist Central Committee in January of 1954, one of the four chiefs of the Yugoslav Government, at times a Minister, head of the Parliament, and Vice-President.

Djilas was born in 1911 in Montenegro, the fateful land he describes poetically in the autobiography of his youth, *Land Without Justice*. At the age of eighteen he went to Belgrade to the University and won early recognition for his poetry and short stories—and notoriety as a revolutionary. He joined the illegal Communist Party in 1932 and was subsequently arrested by the Royal government, tortured, and imprisoned for three years. By the time he was twenty-seven he was a member of the Central Committee of the Party, and in 1940 a member of its Politburo.

Following the German occupation of Yugoslavia in 1941, Djilas became a Partisan leader. In 1944, as a Partisan General he headed a Military Mission to Moscow; the following year, as a Minister in the postwar Tito government, he went again to Moscow to hold talks with Stalin, Molotov, and other Russian leaders. In 1947 he took part in the formation of the Cominform, which had its headquarters, at Stalin's insistence, in Belgrade. In 1948 he once again headed a Yugoslav delegation to Moscow in a futile attempt to stave off the break between the two Communist states that occurred later in the same year.

Ideological disagreements between the Party leadership and Milovan Djilas arose in Yugoslavia beginning in 1953. He wrote articles critical of the bureaucracy he was later to call "the new class," and in January of 1954 he was expelled from the Central Committee. During this period he devoted himself to the writing of *The New Class,* which was to become known the world over for its analysis of Communist oligarchy, and *Land Without Justice.* The year following his break with the Party, 1955, found Djilas being tried and sentenced (a sentence of three years was passed but suspended) for "hostile propaganda" arising from an interview he gave *The New York Times.* After the uprising in Hungary, Djilas criticized the Yugoslav Government's position toward the brutal Soviet action and was, as a result, sentenced to three years in prison. The manuscripts of his two books were, shortly before he was arrested, sent out of Yugoslavia, and the publication of *The New Class* caused him to be brought from prison and, following a third trial, given a further sentence of seven years.

Djilas was conditionally released from Sremska Mitrovica—the very same prison where he had, ironically, suffered as a Communist rebel at the hands of the prewar Royal government—in January of 1961. While in confinement he wrote steadily and he has since completed three books: A massive and scholarly biography of the great Montenegrin prince-poet-priest Njegoš; an historical and fictional account of Montenegro during the First World War; and sixteen short stories (or tales). The present work, *Conversations with Stalin* (in Serbian *Susreti sa Staljinom*), was written during the short period he was free.

On April 7, 1962, Milovan Djilas was rearrested by the Yugoslav authorities, presumably in connection with the then forthcoming publication of *Conversations with Stalin.*